Geography and development

Geography and development

Arthur Morris
University of Glasgow

Routledge
Taylor & Francis Group

LONDON AND NEW YORK

First published in 1998 by UCL Press

Reprinted 2003 by Routledge
11 New Fetter Lane
London, EC4P 4EE

*Routledge is an imprint of the
Taylor & Francis Group*

British Library Cataloguing in Publication Data
A CIP record for this book is available from the British Library.

Library of Congress Cataloging-in-Publication data are available.

ISBNs:
1-85728-080-6 HB
1-85728-081-4 PB

Printed and bound by Antony Rowe Ltd, Eastbourne

Contents

Preface

A state of high development, once so clear and obvious a target for all poor countries and regions, has become more elusive and uncertain. There are no longer the old certainties about what constitutes development, whether it includes such elements as social welfare, quality of life, or education, whether it is concerned with production or consumption, or indeed, whether it is the same for all peoples. Standard goals, imposed by colonial or ex-colonial powers, or through the agency of such bodies as the World Bank, are increasingly questioned by the targets of development programmes, but also by researchers in the developed countries.

Alongside such uncertainties are others: can we actively promote the development of specific regions through the creation of a particular set of settlements, roads, and the opening of particular resources? This was a basis for geographers' interest in development studies in previous decades, but spatial development planning has been an unsuccessful experiment in nearly every case where it has been attempted. More fundamentally, any kind of planning of development seems now, in the neoliberal era, to be suspect, as it can disturb the benevolent influence of free and competitive markets. Intervention to prop up declining industries, or support declining regions, has usually been expensive for central governments, and ultimately has been regarded as a failure or a misappropriation of scarce national funds which should have been used to support strengths, not weaknesses.

Yet another set of fundamental questions has been raised by academics who have drawn attention to the fact that some groups actually resist the development policies of national governments or of any outside agencies. Such cases reflect cultural differences and the conflict of global and local aims, but they also call into question the whole development process. If development is not universally accepted, should it be promoted at all?

In the present work, although different aims and means of development are acknowledged, it is asserted that there is indeed a fundamental underlying unity of purpose. The collective will of people is for progress and for the greater fulfilment of human aspirations. This is not something imposed by the modern era, by the humanist thinking of the Renaissance, or by the eighteenth-century Enlightenment, as has sometimes been proposed, but a reflection of human rationality and abilities for collective action.

Although the book reviews development theories, it also looks at specific cases. Because of what is called the "diversity of development" (van Naerssen et al. 1997) each country has a tale to tell. And although much is now made of globalization, each country has pursued a distinctive path of development with a unique set of institutions. Globalizing efforts of the past – empire expansion by colonial powers such as Britain or Spain – have spawned reactions in programmes of nationalization and attempts at economic autarchy, which have brought their own special results to national and regional development. Globalization is real, but it is not all-embracing and does not mean homogenization.

There are, however, points in common between the development experiences of the less and more developed countries, and both are treated here. The acceptance by the West of the idea of "them and us", stemming from colonial domination of large parts of the globe by Europeans, is not necessarily the best way of looking at the development process, though it is given official endorsement by such agencies as the OECD, the World Bank, and the Interamerican Development Bank.

One important message of this book is that there are some fundamental spatial and temporal patterns of development in modern times. But these are not commonly the patterns on which geographers have traditionally focused, those related to physical resources such as oil and gas, farmland and forests. Instead, key features are now the complex human resources of skills and determination to work amongst an organized and harmoniously living population, often concentrated in dense urban conglomerations where the learning processes are enhanced by a multiplicity of linkages between groups.

There is no good evidence that this will cease to be the case in the near future. The alarm bell of overexpanding populations, or of the final exhaustion of resources, has been sounded often enough, but disasters have commonly been averted by the ingenuity of humankind in finding new resources and in adjusting life styles to new demands. Development, while destroying resources, creates or discovers others which take their place.

Reference: van Naerssen, T., M. Rutten, A. Zoomers 1997. *The diversity of development; essays in honour of Jan Kleinpenning*. Assen: Van Gorcum.

Acknowledgements

Development is a broad field and any text claiming to review it must rely heavily on the work of many others. My primary debt is to the geographers, economists, and others who have done research on the subject over recent decades. Some of these have worked in the countries of what used to be called the Third World and have attempted to analyze projects and the actual course of development in these countries. They have had the difficult task of trying to interpret local events in terms of aims and models established for other countries, often in the absence of good data sources.

Locally, I am indebted to my colleagues, at the University of Glasgow, for their support in maintaining a research-friendly environment. The students who have chosen my classes in the Geography Department have also helped over a number of years, either by providing a sounding board, or by questioning the established wisdom on any subject. On the technical side, the departmental secretaries were a valuable resource, and Mike Shand was able to produce maps at short notice from rough materials.

In London, Roger Jones of UCL Press was diligent in pursuing me for completion of the work and, after his departure, Steven Gerrard and the editorial assistants were helpful in keeping the momentum going.

CHAPTER 1

The nature of development

The drive for improvement in human conditions, in personal happiness and in social wellbeing, is innate in every society. It has been more openly expressed in the last 50 years, and that part of happiness which lies under social rather than individual control has been the object of endeavour by theorists and public agencies seeking to understand the processes involved and to control them. At the broadest level, these aspects of human improvement comprise what is involved in development.

Fifty years ago the term "development" was used largely in the context of economic change. Economic growth may be defined as the increase in production or consumption of a nation or a region, while economic development is the increase of such production or consumption by each person, putting growth onto a per capita basis. Economic growth may increase the weight of a nation in world affairs, but it may fail to make life any easier for its inhabitants. Economic development provides this increase in goods and services which may be felt by the population.

The geographic scale

As understanding of some of the complex links between economy and other human processes has increased, so the definition of development has been amplified from the purely economic to include other elements that lie within the scope of social action. The subject of development is also studied more widely, by economists still but also by sociologists, political scientists and geographers as well as other social scientists. In this work, we focus on the spatial aspects, and particularly at the regional, subnational level, although some of the observations concern whole countries, since these too may form geographical patterns. Regions that form large and economically, socially and culturally identifiable parts of a nation state are the obvious focus, because they have been the object of most analysis regarding spatial differences in development levels or wealth, and the object of policies which individual governments have put in place to modify these spatial differences. Differences within countries should be amenable to change through policy because this can be controlled by a single state; it is also seen as desirable to reduce these differences, for reasons of bringing common

1

levels of equity to all of a country's people, and from the point of view of individual governments, to stem the possible rise of regionalist unrest which may otherwise grow into local nationalism and the move for separation from the mother state.

At certain points in the discussion it will be necessary, however, to look at a broader scale, that of whole nations. At this level, there are patterns, as for example in the case of East Asia, where a common thread of development processes seems to affect a whole set of countries following in the lead of Japan, and there is a perceptible spatial diffusion outwards from the Japanese core. Yet other kinds of regions, some no larger than the intra-national regions that are the usual object of study, can also be observed, as national boundaries are traversed in the development of some border regions, such as that between the USA and Mexico. Other regional groupings are those of key centres like Singapore and Hong Kong with industrial nodes at some distance from them, in mainland China or Indonesia. All these are economic regions that have defied the boundaries of the nation state, and in Chapter 3 we consider whether they are becoming more important than the regions within states.

The lost consensus on the development process

In the 1960s, it was seen as quite clear that economic development was a desirable aim for all countries, and that it was to be achieved by a set of standard measures, usually including a broad industrial plan to raise a country's level of production, its exports, and the diversity of products. It was also accepted that in the course of development, some regions of countries would lag behind, and that this was undesirable. Intervention by the state to assist in overcoming the differences that might build up was thought to be the best solution to this problem. A process of development was seen to include changes from economic activity concentrated in the primary sector (farming, fishing, forestry and mining) to a dominance of secondary activity (manufacturing), and then, in the later stages, to dominantly tertiary (services) or quaternary (research and development) activity. A spatial process was also identified, development (or at least technological change) starting in key cities and spreading out from them down the urban hierarchy and into rural areas. This process could be called a diffusion of development.

This diagnosis is no longer acceptable to many. On the one hand, the concept of economic development has itself broadened out to include people's social and environmental aims; different goals for different regions or countries are recognized; and the role of the state or of planners is questioned. On top of this the diffusionist model is doubted, and the accepted role of geographers in preparing the spatial diffusion models is undermined by the failure of these models in real life. It might be thought that these new uncertainties of the 1990s make it too dangerous or irrelevant to write about development from the geographer's point of view. But this book is intended to show that there are numerous aspects on which geographers have much to offer. For one thing, they are able, perhaps more than economists or ecologists or sociologists, to consider a large set of aims for development, including social and environmental aims alongside the economic ones. On the question of state involvement, geographers have a large range of examples from different countries to demonstrate the role of state versus individuals and other institutions. Finally, there remain important spatial aspects to development, even though the diffusionist model may be shown to be untenable or irrelevant to many specific cases.

The nature of development

In the 1940s and 1950s development was identified as a purely economic process. In fact, the term consistently used was "economic development", leaving aside any discussion of social, personal and other aspects. In the literature on the subject, most writings came from applied economists whose concern was for the betterment of production systems in the countries devastated by the Second World War, and thereafter in the world's poor countries, then identified often as the "developing countries" (indicating that they were moving forwards), the "underdeveloped countries" (with the idea of perhaps a more permanent state of affairs), or increasingly commonly, as the "Third World", referring to a political status outside the great power struggle between East and West, capitalism and communism, and identified by poverty.

In the late 1960s, development as a concept came increasingly to mean socioeconomic development, which meant that the old measure, gross national product (GNP) per capita, was inadequate to represent it. A complex of measures showing levels of education (more frequently of educational provision, which subtly alters the idea), medical aid, housing, and other goods and services which entered loosely into the concept of welfare, came into use. Since at that time, the height of the "welfare state", many of these elements were provided publicly in advanced countries, welfare measures inevitably emphasized the higher status of these countries, and at the same time made the case for more state involvement in development, to provide more welfare.

What might be termed a third phase of thinking in the 1980s has come with the further spread of the development concept, to include aspects of the environment such as freedom from pollution, whether atmospheric, noise, water, or visual landscape pollution. Access to countryside and the ability to enjoy a healthy lifestyle have been included in measures of quality of life, which also includes the older economic and social elements such as income levels, and welfare provision such as schooling and medical aid.

Many of these quality-of-life elements are of low importance to people in poor countries. First needs, or basic needs as they are termed in the literature, are for shelter, food and clothing. Beyond these, once they are satisfied, are the welfare elements; and beyond these again are the aspects of desirable environment, unpolluted air, or community spirit (Maslow 1954). What is observable is thus an evolution of needs, which matches the evolution of academic definitions of what development is about. For the advanced countries such as the USA, the environmental features and sense of community are likely to be valued very highly, as food and clothing are effectively taken for granted. In Brazil or Bangladesh, food and clothing are likely to feature as important aims for a significant segment of society. In the USA, states and cities could be and were compared in terms of their provision of a broad range of facilities which go to make up the quality of life (Cutter 1986). There are comparable studies of British cities done in the 1980s, where the numerous elements involved were weighted by consumer indications of their importance (Rogerson et al. 1989). Clearly the list of items here could be extended indefinitely, to include personal states of mind which also contribute to human happiness. In the geographical and sociological literature on the subject, however, attention has been confined to those elements that are available publicly, not individual matters which come into the realm of psychology. In common with the welfare geography emphasis, quality-of-life studies have generally concentrated on consumption aspects, which begs a question. It is always necessary, for city, region or country, to produce the conditions that allow consumption. Thus in recent decades, the attention of academics and planners may be said to have been somewhat diverted, by failing to concentrate sufficiently on the supply factors, resources, technology and human organization which allow development.

Processes

Apart from what is defined as part of development, briefly reviewed above, there is the question of what kinds of processes are involved. Most students would identify many interrelated processes, but the most important are likely to be: the growth of per capita income, whether measured as production or consumption; a series of demographic changes towards a modern society with small families and low death and birth rates; and a set of unbalancing features, including the concentration of population in a few cities, of industries, and rising inequalities between sectors of society and between regions of countries.

This view of the development process may be enlivened and given a time sequence by considering what Alonso (1980) identified as a collection of parallel evolutions, his "five bell shapes in development". In a period of development, he observed, there are several features which seem to move in the same way, rising to a peak and then subsiding. They are: (a) the economic growth rate; (b) the level of social inequality (between classes, occupations, races); (c) the level of regional inequality; (d) the level of geographic or spatial concentration (urban-industrial growth in a few big cities); and (e) the population growth rate, in demographic transition (from high birth and death rates to low birth and death rates). It is important to note that these five bell shapes do not occur at the same time, and that they are closely related, although the relationships will vary between countries and regions. Alonso's impression was that geographic concentration is a first process, followed by economic growth and then by social and regional disparities over time.

In reality, the demographic transition may be posited as happening in radically different ways in advanced and poor countries, starting early in the poor countries because medical aid to death control arrived, from the exterior, before the rest of development, but finishing late as people failed to adopt Western cultural norms about family size. In the advanced countries, medical aid was only one of a number of technological improvements which occurred alongside one another, and this problem of separation did not occur.

On top of Alonso's bell shapes, there are the changing concerns of people mentioned earlier, shifting from such survival elements as food and clothing, to welfare and then to quality of life. Putting these into the frame, there is a broader picture of change. In traditional societies, the population and economy are dispersed, and dominant concerns are those of survival. In the next stage, of early development, there is the process of concentration of economic activity and population in towns and cities, accompanied by a shift to welfare concerns, since the new urban society does not provide mutual communal care as it did when people lived in small communities.

Awareness of social inequality and regional inequality grows over time as people become more educated and have the chance to observe other groups and other regions. Political parties, movements, and regionalisms or local nationalisms, may grow out of these concerns in a third stage. This will often provoke regional development policies as well as policies for the reduction of poverty or of differences between social groups.

In a later stage, the concerns move on to quality of life, and these concerns may be expressed in a reversal of the concentrations of industry and population, as technology permits decentralization and the movement of people out to more attractive rural surroundings. Counterurbanization, a move away from the big cities in a demographic and economic activity sense, is a feature of most West European countries, as it is of the USA. In some countries, such as the UK, this stage may be reached when there are still major concerns over regional or sectoral inequalities, so that the development drive exhibits a complex set of different aims for different groups in society. Thus Scotland, for example, presents strong local nationalism linked to cultural past as well as to the more recent industrial past. At the same time it is the recipient of inmigrants from England who are leaving the large cities and seeking better quality of life.

5

Space and time

Over the years, more sophistication has come to the understanding of the temporal and spatial processes of development. Two points may be made here which will be returned to again at various points in the discussion: development occurs in a lumpy, irregular form, first in time, and secondly in space. On the time dimension, W.W. Rostow (1960) set up a famous approximation to the development process by asserting that it could be compared to the take-off and ascending flight of an aeroplane. An even progression was portrayed, from preparation for take-off (among the poorest countries), taxiing for take-off, take-off itself, and steep climb until cruising height (the mature, high-consumption economy) was reached. This idea, coming from an analyst of economic history and based largely on European experience, was soon questioned, and the dependency writers in particular claimed that development could not only be stalled, but even put into reverse, perhaps with a crash finish, although the aeroplane analogy was not pursued!

Another version of the irregularity over time has been in terms of waves. At various times, and more insistently in recent years, the concept has been advanced that development may be visualized in terms of waves of long duration, during which a particular technology is the lead and a number of products are associated with it. After the burst of development associated with one wave, there is a period of stagnation or retreat before the next. Kondratiev waves, as they are named, after the leading Russian exponent of the idea in the early twentieth century, contribute much to our understanding of development (Hall & Preston 1988). Another set of ideas which links in with this view is that of Schumpeter (1939), who showed how innovation linked to business. With a new technology, there is a rise in employment and income and a positive attitude towards ongoing research and innovation. As the products mature, there is less margin for improvement and competition reduces profitability, and many firms cease innovating. Replacing human labour with machines reduces incomes further and there are job losses. Finally, innovators perceive new opportunities, and people without jobs are very willing to take a risk and move on to a new technology. Each of the technologies uses new human and physical resources, so that development 'hits' on one or several regions, stays with it for a while, then moves, often quite quickly, to focus on a new region or regions with the best factor mix.

Regional development in Britain will be described in this book in terms of these waves (see Ch. 6), although it is shown in the same chapter that the equivalent process in Spain, while it is also an uneven process over space and time, does not correspond to this model. For the less developed countries, the process of development is so recent that the waves may be visualized as being virtually compressed into one. The general point, however, is the irregularity of development impulses over time and space.

A second point is that development is not a spaceless phenomenon, and it also occurs in lumpy form over geographic space. Specifically, it tends to be highly concentrated, especially, over the last 200 years, in urban-industrial nodes.

Through a series of positive feedback loops, these nodes tend to be self-maintaining over a period. New industries are attracted to them because of a variety of advantages in provision of goods, information and services. In some cases, a diffusion outwards of their dynamism may occur, and in others the impulse of development is retained in a single centre, but the central fact is the clustering and cumulative nature of development changes: in geographic space, and in the abstract spaces of the financial and business worlds, allied to their clustering in time. This was the central point made by the studies of François Perroux, although his work was reinterpreted into a planning tool of dubious value by later workers (Perroux 1988). Concentration is thus a feature which is unavoidable, a point that needs to be borne in mind when we consider the various efforts of policy-makers to redistribute wealth or production in regional policies.

Another point which may be made about the spatial or regional effects of economic change is that each region, being smaller than the nation it belongs to, is liable to feel the effects of change more drastically, because in general it will have a narrower economy, relying perhaps on one important product, which itself comes from one or two major local resources. Each region has an economy, in the modern world, which is more fragile than the nation, and policies for helping development, if they are needed at all, are perhaps most needed at the local or regional level.

Anti-development

As economic change under capitalism has gone on since at least the time of the Industrial Revolution in the late eighteenth century, there have been a series of revolts against innovations. These varied from those attacking the introduction of machinery or new products (like the Luddites, a group of English workers who destroyed the textile machines that they rightly saw as destroying their own more primitive livelihoods, in a movement which ran from 1811 to 1814), to Utopians like Robert Owen, who sought the alleviation of poverty in the early nineteenth century through the creation of new communities on the frontier where they could reform society; and forward to the modern movements against standard development practice, such as the advocates of Development from Below, or the new environmental movements which ally protest against capitalism with concern for the environment.

At their most pessimistic, the protesters claim that modern changes, such as the liberalization of the economy being undertaken in Africa, bring negative development and impoverishment to large sectors of society, and particularly to the rural peasant groups. While these protest movements are well founded in reality for rural localities, they often bypass the main changes taking place in the economy to which they refer. Rural regression in Africa (Taylor & Mackenzie 1992) does not preclude urban advance in the same continent, and the advance of

7

the urban may be the salvation of the rural. In cities, social customs change with rapidity, such as norms regarding the size of the family. Thus for demographic change, to limit the tide of population growth that threatens to overwhelm countries such as China and India, migration to the cities, rather than a re-arrangement of rural life itself, may be the main part of the solution. It is thus necessary to take as wide a view as possible of the economy and society to which development processes refer, since features at one level are sometimes compensated by those at another.

It is also necessary to realize that change or transition from any technology to another will involve hardship and upset to given ways of life. Steam engines and weaving machines displaced cottage workers in nineteenth century Britain. Today, computers are displacing workers throughout the developed world, while "real jobs" making things such as car components are exported to distant countries. But social and economic change means that other employment is always found for the displaced, and the transitions are taking place with increasing rapidity.

In addition to the non-acceptance of development by groups in particular regions or countries, especially those who are displaced by the changing course of development, anti-development is a feature of academic discourse. Texts on the subject over the past 50 years have generally accepted the idea that development is desired by human populations and that it is indeed desirable. This has had the effect of reducing or even eliminating any discussion of whether what the developed countries consider to be development actually conforms to the interests of the less developed countries. A feature of the late 1980s and 1990s, however, has been a fundamental critique which asks whether development is an agreed aim, or whether it might be an academic or institutional set of ideas and practices with only limited contact with reality. The central arguments here are borrowed from French linguistic philosophers who regard many of the academic bodies of theory as simply that: theories which are not based on an objective reality but on an internal logic all their own.

One line of argument of these writers, who belong generally to the post-structuralist school, is that development is an idea created as a solution to a proposed problem, that of poverty in the less developed countries (LDCs). Poverty was, according to this interpretation, only "discovered" after the Second World War, and to solve it there has come into being development, which is essentially a technical matter (Escobar 1995, Ch. 2). Going on from this conclusion, the West has experts and institutions ready to help solve the problem, and the end result is intervention in the poor countries of the world. Overall, the argument is that the need for development was invented to justify the grand world strategies of the major powers.

Another argument in the same vein is that development is a replacement of, and a logical continuation from, colonialism. As colonial power declined in the 1940s, development ideas were instated in its place, and power for the developed countries to intervene was maintained. Ideas on the need for development were discussed in the 1930s in the British Empire, and these fitted smoothly into the post-war ideas for economic development.

In more moderate versions, the arguments are not wholly against development, but for "another development", picking up a variety of themes which basically oppose the top-down organization of development efforts. Ecodevelopment, self-reliant development, ethnodevelopment, basic needs, and a challenge to the market rules of capitalist society, are amongst these themes (Hettne 1990). Much of the field had been covered within the Development from Below ideas of the early 1980s, but it has resurfaced and has been reinvigorated by the continued failure of standard efforts in development to reach some groups of society, notably in the poorer countries.

Much of the work in this anti-development literature is concerned with rural, peasant or indigenous groups, at one extreme, and with international relations, especially the matter of power, at the other extreme. This does leave out some of the main ground for discussion of modern development, which is very much focused on urban and industrial themes. This point will be taken up in the next chapter. It is also the case that much of the argument is not about the majority groups of the population, but minorities: ethnic, regional, peripheral or in some other way disadvantaged. It must be allowed that these groups are likely to be opposed to development, because of their collective cultural differences, without detracting from the arguments for a positive value for the main segment of the population. Another 'minority' group, without a proper voice for itself in an economically regulated world, is seen as the environment, and the environment is seen as defended by the disadvantaged groups rather than by any developmentalists.

Development and the environment

An important feature of some of the discussion regarding individual countries in this book will be that development may be either fostered or checked by the attractions of resources, often located in large but remote peripheral regions. In the next chapter, the old thesis of resource-led development, the staple or export base theory, will be shown to have little relevance in the twentieth century. In a world with globally available raw materials and power sources, the focus of development has been on concentrated industrial processes and services in cities, and those countries with the best development record over recent decades have had little reliance on their own resources. The East Asian 'tiger' economies may here be contrasted with those of South America.

The relationship of resources and large regions to development is, of course, not a simple and direct one, but geographers should note the lack of a direct relationship, and indeed the apparently perverse situation where one of the classic factors of production, the land, fails to support strong development. As is detailed

9

with regard to Latin American countries, the secret of this failure relates to government policies attempting economic autarchy, and policies promising the development of particular regions with political or geopolitical importance.

Environmental deterioration is an important issue with respect to development around the world. There is little question about the reality of this deterioration, although rates of soil erosion and pollution are not wholly agreed. What is more in doubt is the location of responsibility, and the different actors who are supposed to be the culprits. From the previous paragraphs, it might be concluded that national governments are the most guilty because of policies for development of remote regions without any care for conservation. Geopolitics is seen as a major interest in Brazil and all the Amazonian countries, in their programmes for deforestation and settlement of the rainforest.

There is a contrasting argument that states that capitalism and external, colonial or neocolonial powers are the main culprits (Redclift 1987, Gligo 1993). Because outsiders (the colonizers of empire in the past, or the multinational corporations of today) are not interested in conservation but exploitation, they destroy nature. They also have broken down traditional societies which had a good understanding of the environment. People have been replaced by machines, land which was communally owned has been expropriated by new landowners, and the most experienced people have migrated to the cities. This has left rural people with only remnants of indigenous knowledge, and often with insufficient land resources to be able to conserve them, instead being forced to exploit their own soil to the maximum.

Both these arguments are reasonable and worth discussing in the LDCs. They also imply different kinds of solution. If national governments are to blame, the political solution is through the advent of democracy with a strong voice for those interested in maintaining the environment. If external forces are to blame, governments might seek to close themselves off from the exterior, or at least to control more closely the actions of foreign firms. Decentralization and the implementation of local and regional plans are also supported by the realization of special environmental problems. One of the most widely accepted conclusions of the Development from Below ideas was that people local to any one region are those most concerned with their environment, and are more aware of any deleterious effects of current management.

Positive and normative approaches

There has commonly been no clear division between development accounts that state what has been happening in the real world and what should happen: what the jargon of economics distinguishes as 'positive' versus 'normative' approaches. For example, while a writer such as François Perroux wrote on growth poles as a matter of positive economics, later interpretations of his work were largely normative, in terms of policy. Because of this mixture, it is impossible to treat the subject now in a broad review without referring to both sides. However, an effort will be made in subsequent pages to separate clearly between normative and positive aspects.

Normative aspects come to the fore when we consider regional planning and policies. Plans for regions and for nations take a view of what is desirable and how society should achieve the desirable goal. Just as there are disagreements as to the nature of the development process, there are a variety of stated goals and policies which might achieve these goals.

One of the principal goals of spatial or regional policy over the years has been the reduction of inequalities between regions. Great inequality is undesirable because of the frictions or tensions created between different social or regional groups; but action to reduce inequity is usually at the expense of efficiency. For example, if the efficient level of steel production in a country is with two very large mills located near major markets, it is inefficient to have four smaller mills, some of which are at a distance from their main markets. But this was the regional policy adopted in Britain for the nationalized steel industry of the 1950s. Equity versus efficiency is thus a real question in the context of this kind of regional policy. Alternative regional policies might ignore equity or ignore efficiency.

That there can be other goals may be briefly summarized. One such goal is now seen as sustainability, the pursuit of policies that maintain resource levels for the future. Another, not normally stated directly, is the policy of some states, such as those of Latin America in recent decades, to achieve development of peripheral regions for the sake of geopolitical goals. This has been the case in large countries with thinly peopled borders, such as Brazil.

CHAPTER 2
Theories and models

The debate

At their simplest, the basic theories on regional development can be portrayed as a left-wing versus a right-wing view of development. The right-wing view (also termed "liberal" or "orthodox") is that regional inequalities arise through the operation of normal economic forces in the course of capitalist development, and disappear also through the working out of these same economic forces. Left-wing views (also "radical", "structuralist") also have regional inequalities as coming about through the operation of capitalism, but interpret these differences as being instrumental to the maintenance of capitalism, and actively promoted by capitalist agents, whether firms, individuals or governments. Because of their importance, the inequalities between regions are maintained by capitalist powers and do not disappear through a process of convergence.

These two sets of ideas were present from the early writings after the Second World War, and are epitomized in the opposition of views between Albert Hirschman and Gunnar Myrdal in the 1950s. Hirschman (1958) envisaged a spread of development out from centres, and Myrdal (1957) concentrated instead on forces which would tend to maintain or increase differences between regions.

In the 1970s and 1980s, following this debate, other lines of thinking came into fashion as it became apparent that neither left- nor right-wing analyses were able to be translated into practical policies. Evidence accumulated that in many parts of the world there were widening gaps of income and welfare both at an international level, between different countries, and at an interregional level, within individual countries. Ecodevelopment, Development from Below, sustainable development and basic needs are some of the concepts that have been discussed in this debate. In what follows we may generalize to call this the "endogenous" school of thought, reflecting sources of the development impulse.

While there are many distinctive aspects of each of the concepts within endogenous development, one common thread is a critique of the idea that development has to come from somewhere else, that it is exogenous or from an outside source. Instead there is the concept of local, endogenous development based on local resources, both human and physical. In Riddell's (1981) ecodevelopment for example, there is a stress on the need for local control over resources in order to conserve them and not deplete them. In Development from Below (Stöhr & Taylor 1981), the emphasis is on the use of local resources by the local population, and this concept is stressed further in the agropolitan development model of Friedmann (Stöhr & Taylor 1981). In all versions of this kind of writing, local control over human and physical resources is advocated, through decentralization of power to the local level. Local initiative is also promoted, with the associated ideas that local initiatives will lead to local entrepreneurship in projects and programmes for development. Local initiative will also link to the use of local, perhaps intermediate, technology, as well as local resources rather than those that have to be imported. In general, integration of the different sectors of the economy will be fostered by such an approach. For example, a local crop, say sugar cane, might be processed into local industrial goods, such as sugar or cane alcohol, and the waste bagasse might be sold as fuel for other local industries.

Some of the ideas were generalized by Friedmann & Weaver, in their book on *Territory and function* (1979). Here the contrast was drawn between one kind of development of regions, territorial development by and for the region itself, and another kind, functional development, which is development of the region for the nation. Using the case of the Tennessee Valley Authority (TVA) it was shown how this agency was set up by central federal power in the USA, but was run very much in the Tennessee Valley in the interests of the local, regional economy. Alongside efforts to improve river navigation and to increase electric power output in the valley, a major effort was made to improve the lot of small farmers, those producing maize on steeply sloping land as a monoculture, and therefore responsible for the central problem of soil erosion and river flow instability. Extension work on improvement of farm management techniques, diversification of production, farmer education, and infrastructure improvements such as farm-to-market roads, were all part of the programme in the 1930s. In the post-war era this changed, as the federal government saw the possibility of using the region as a major power generator. Alongside the older hydroelectric schemes there came into production nuclear generators, and the region became a source of cheap power for the nation. In every country there is the possibility for this kind of functional approach to replace territorial concerns in regional development.

Within the endogenous school of thought there is an obvious extension of the aims of economic development, to include aspects of welfare such as provision of housing and education, and the quality-of-life elements, such as maintenance of an attractive environment, freedom from pollution, and existence of a strong community life. These were not previously counted alongside the economic production aims. However, once local concerns and priorities are placed first, or at least included alongside others, then various non-economic interests will come to the fore.

Summarizing the above discussion, two strong lines of argument, which we have labelled "left" and "right" versions of development, held sway in the first decades after the Second World War. However, data from observations of national or regional level development do not provide a firm base for assertions about a "normal" course of development, nor for the correctness of either general view. We may briefly reconstruct some of the main tenets of theory and indicate areas of similarity or agreement between the two sides, as this may lead to consideration of what constitutes a real alternative. One important line of argument arising in the 1970s and 1980s was is that neither left nor right were correct about the sources of development, and that external factors were overemphasized. Instead, the emphasis should be on internal or endogenous sources of growth and development. Such a view comes from concepts such as Development from Below and sustainable development. This idea implies that development priorities are likely to move from the purely economic towards the inclusion of social and environmental concerns, and to involve a measure of decentralization of power to lower administrative levels.

The neo-classical model of development

As a representative of the right wing or orthodox view of development processes, we may take the neo-classical model of economics. Without entering deeply into the economic arguments, this may be described as a model of balance, in which any slight imbalances are self-righting, so that no problem of inequality should exist. This result is achieved partly because of the many simplifying assumptions in the model (Richardson 1973). The basic model of production for a country or region is $Y = aK + bL$, which may be regarded as meaning that production (Y) is equal to some constant mix of capital and labour. The three classic agents or factors of production are taken as being capital (K), labour (L), and land, in which latter term all natural resources are included. Since "land" is considered to be fixed, it can be left aside. The constants a and b in the equation refer to the fixed amounts which are necessary for each of the two factors considered. One implication of this simple model, since a and b are set for the area under consideration, is that there is a fixed ratio between capital and labour, the ratio $K:L$. For the purposes of illustration, let us suppose that the ratio is 1:1. A further implication is that any emergent imbalance will be righted towards the initial balance, by a migration of the overabundant factor to the region where it is scarce. In Figure 2.1 we may visualize a sequence of events in a two-region country; in stage I, the $K:L$ ratio is equal in both regions. In stage II, the ratio is disturbed by san investment from region B into region A of 10 units. (We might suppose a firm has set up a new factory in region A, and this is a common form of capital movement.) In stage III, the balance is restored by a movement of labour from B into A (i.e. a migration).

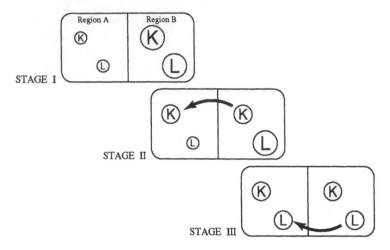

Figure 2.1   Neo-classical model of regional developments.

Some criticisms

There are obvious limitations of such a model. Its suppositions are an extremely limited world, where most factors are held constant or assumed away. In reality, it is a model that refers to one product, not to a whole regional or national economy, and diversity of products introduces great complication. The markets are assumed to consume all the production and any increase in it. In the simplest version of the model, there is no technological factor, i.e. no advance of one region over another because of new inventions or the use of better machines. It is not a development model either, but one for economic growth, and development, involving some increase in production per capita, would depend on other factors such as technology.

Capital movement

From the point of view of geographers, there are grave defects in the assumption of zero transport costs for capital and labour. Taking capital first, most people think of this as financial capital, money in one form or another, whereas in reality most capital movement is through investment in physical structures such as roads and factories. Even if it is financial capital, flows may not be easy as there are government restrictions on international bank flows to and from many countries. Taxes and risks of capital loss through devaluation are also variables which are considered by those moving money around the world. In the late 1970s and the

1980s, there were indeed large money flows around the world, as oil money from high oil prices sought placement, often in developing countries, through banks and other institutions in the developed world. But this process has probably been an exception rather than the normal way in which capital moves.

In the case of major flows which are to have an impact on a country's development, most of the relevant flows are not money movements, but what is called fixed investments, i.e. capital put into building factories or roads in order to produce goods in the target country. Such investments are "lumpy" (i.e. they come in large amounts or not at all) and firms study carefully where and how the investment will be made. This kind of investment is also highly visible in the object country, so that political considerations may come into play. In countries with powerful nationalist movements, national government may find it is undesirable to bring in foreign firms which wish to show off their products or technology, and even their higher wage levels may be an object of criticism by local firms which cannot compete. In the countries of Latin America, inward investment in factories was spurred by high tariff barriers, taxes on imported goods which were so high as to make it advantageous to set up a local factory in the country and bring in parts to be assembled there, even if this was at high cost.

Labour migration

Similarly, labour movement is not a simple economic process of movement of individual workers to high-wage areas. Instead, labourers belong to families and the whole family is involved in decision-making. Families may decide not to migrate because of their sentimental attachment to a home village or ownership of land there. A positive decision to migrate may be driven by social factors (contacts in the city, family friends, friends to help with employment or accommodation) more than economic ones. It would also appear that much migration is driven, not by present wage differences, but by the long-term probability of improving the family situation through education and moving up the employment ladder. Information is vital to the decision to migrate. Big cities become the target of much modern migration, partly because these are the cities that potential migrants have heard about, via radio, television and other sources of information.

Other criticisms were made over the years, including the concentration of the neo-classical theory on production, and its failure to consider distributional problems: in other words, was the production being consumed by the population at large or by a small elite (Friedmann 1988, Ch. 2)? There was also a historical critique, to say that the development paths of the poor countries were not and could not be like those of the now developed countries. W.W. Rostow's model of development could not be transferred straight from eighteenth century western Europe to the rest of the world.

Spatial versions

There is no precise "geographers' version" of the neo-classical model, nor can there be, as transport costs are assumed to be negligible, contrary to all the emphasis in geography. But an interpretation of neo-classical thinking into a spatial structure can be made through adding some "spatial friction", while maintaining the idea of factor flows which have a broad balancing effect between regions

Centre and periphery

Two writers described most fully how neo-classical ideas can be translated into a spatial model. Williamson (1965) described the way in which interregional disparities might change over time through a developmental process, with statistical evidence from many countries. This is a time–space model. Separately, Friedmann (1966) described the evolution of these differences in a hypothetical country where flows took place between the centre and periphery.

Williamson's contribution was to show how countries change over time in their degree of income variation between regions. In an early stage, interregional differences were thought to be small. Over time, as the first development impulse was felt at one point, perhaps a port city or the main industrial centre, differences grew (measured in per capita income) between the regions, because the other regions, perhaps rural regions of subsistence agriculture, failed to change. Differences reached a high point at some intermediate level of development, when presumably the central region was advancing fast, but most regions remained at a subsistence level, and thereafter the differences in income reduced, from the evidence of the most advanced countries, to very low levels again. This model of change was not based on historical data of the actual changes in one country over time, but on data from many countries at different levels of development, from which the actual process could only be inferred. Only for the USA was there actual data showing this process to have happened over time, which means there could be major flaws in the argument. Specifically, we do not have evidence that India, at the bottom end of the development curve, will become like the USA is today in terms of its regional balance. Possibly, no country today can expect to follow the advanced countries because these latter have a head-start in industry and others must follow a different path. Some fragmentary evidence for LDCs is that the differences between regions are not diminishing but widening in the present-day global economy. This criticism is the same as that levelled at W.W. Rostow's (1960) famous model of economic development. Rostow's model was in the same vein as that described by Williamson, and is portrayed as a linear progression in one direction, using the analogy of an aircraft taxiing and making progress through take-off and climb, to reach a high cruising height when fully developed. There is insufficient evidence to show that countries do all "take off" and climb towards this high level.

One well-known spatial model based loosely on neo-classical ideas is that of Friedmann (1966), who described a centre and a differentiated periphery, in which the movement of factors of production, labour and capital were portrayed as following the predictions of the model. Friedmann's model hypothesized a first development in one central point, the largest or most important city, which leads to an imbalance between the centre and the periphery. The centre sets up the first manufacturing industry and accumulates wealth, which affects the build-up of services and administration in the central city. Over time, however, there is migration of factors, and Friedmann postulated the outward movement of some capital – in the form of factories which are too confined in the centre and seek cheaper land and labour in the outer regions – and the inward movement of labour as migration from rural areas. These various movements are benign, in that they generally tend to reduce the inequalities of income between regions. A need for some governmental intervention is seen, however, especially in the intermediate periods when the greatest differences between centre and periphery are seen.

The set of movements defines four kinds of region according to this model: the centre, with its rapid growth and problems of congestion; the "upward transitional" region, with capital in-migration as the recipient of overspill factories and also immigration of people; the "downward transitional" region, with out-migration of people and capital; and a fourth kind of region, the "resource frontier", with slight population but massive natural resources. This final kind of region reflects the fact that Friedmann elaborated the model with reference to South America and especially Venezuela; elsewhere, such regions are obviously not always found. In later years Friedmann reversed his view of development by joining the dependency school (1973). His original view, however, presents a clear case for the diffusionist arguments.

Attached to this centre–periphery thinking, and to the neo-classical vision of development, is a literature concerning balanced and unbalanced growth, which was the central debate in development theory of the 1950s (see Agarwala & Singh 1958 for review). On one side, an early view was that balanced development was needed, especially balanced in the aid being given to help the revitalization of the European economies after the Second World War. The argument here was that these economies had suffered in all sectors, and failure to help any one would create bottlenecks that would thwart development. The contrary view, which came to prevail, was given by Hirschman (1958), who advocated unbalanced development as a necessity in aid programmes, on several grounds. One was that this was the historical way in which development had always happened – one sector had the lead, and others had followed. For example, the railways had been a nineteenth century lead sector, and attracted all the best brains of the day, as well as the capital available for risky ventures. Other sectors had been brought along in the wake of this dynamic sector. A related argument was that each country had a limited amount of skilled people and administrative ability, so that a focused approach was imperative. Hirschman's ideas, on the need to concentrate, combined well with his view that development did diffuse out to the periphery: the diffusionist view which contrasted with Myrdal's and others' espousal of the dependency arguments.

Growth poles

Growth and development in a centre, spreading out to other areas over time and space, are also concepts contained in the earlier ideas of the French economist François Perroux, who studied the idea of a growth pole. His work is mostly known through its adaptation to spatial planning, but his own work was mostly to show that developmental impulses are concentrated. For him, development was led by a growth pole industry or sector. Its three central characteristics would be its dynamism through constantly new technology; its rapid growth as an industry; and its wide linkages both horizontally, to other industries round about it, and vertically to suppliers of raw materials and to markets for its products.

Once such a view had been propounded, it was readily translated into a geographical space, so that a growth pole could be defined on the ground, as a place with growth pole industries having the characteristics of growth, innovation and linkage. A growth pole could be a "naturally occurring" industrial city, or the idea could be used as a planning tool, since growth poles could be created to induce growth in a whole region. Through the multiple links of such an industry, it would encourage all kinds of other industries in a kind of halo around it.

Further extensions of the growth pole idea would link it to the whole urban structure, making growth poles of the intermediate cities to act as regional stimuli outside the central one. A theory of central places had already been elaborated by geographers, with the observation of a general regularity in the urban hierarchy, with few larger centres, and at different lower levels, many more centres offering lower order services. Looking at real-world urban hierarchies, it could be seen that many countries, like post-war France and Spain, had one or two large metropolitan centres, but few intermediate cities acting as regional centres in a second level of the hierarchy. Following on from this reasoning, growth poles could be created through the expansion of existing intermediate cities, so as to have a regular hierarchy which would be better able to distribute growth down from the centre, as well as stimulate the region around it.

The French example

A major testbed of the growth pole idea was France, where it might be argued that the best chances of success for a spatial application were found. As early as 1947, Jean François Gravier had criticized the regional structure of the country in a report entitled *Paris et le desert français*, the title revealing how the country had become overcentralized since pre-Revolutionary time and continued to focus all economic activity and administration in Paris (Gravier 1970). Within France

(Boudeville 1966), the regional planning of the Fifth National Plan used a spatial version of Perroux's ideas to create eight *métropoles d'équilibre*, regional cities or pairs of cities which would act as counterbalances against the centrality of Paris, and located well away from the centre. These were at Marseilles–Aix in the lower Rhône valley, Lyons–St Etienne in the middle Rhône, Strasbourg in Alsace, Nancy–Metz in Lorraine, Lille–Roubaix–Tourcoing in the north, Nantes–St Nazaire on the Loire, and Bordeaux and Toulouse as two separate centres in the southwest. Following the metropoles, other plans were for industrial poles focusing on heavy industries, at Lyons, Paris and three coastal sites (Le Havre, Dunkirk and Fos sur Mer). Later modifications of the policies took them down to smaller towns in more rural settings.

It is difficult to assess the success or failure of the growth poles policy in France. There has been a major element of decentralization of economic activity from Paris over the decades since the 1960s, but it is impossible to say how much of this has been because of the growth poles and their successor policies. Some individual poles must also be regarded as failures or partial failures. One has been the industrial pole at Fos sur Mer, supposedly associated with the Marseilles–Aix axis of development (Bleitrach & Chenu 1982). This huge development, of a steelworks and associated oil refineries, petrochemical and aluminium works, was proposed in order to generate industrialization and economic development in the south of the country. It is criticized on two grounds. In the first place, it is regarded as having been merely an agent of central government in league with the dominant capitalist interests of the big banks and steel companies. Government money was put into Fos which simply subsidized moves that would eventually have been taken by the companies themselves. In pre-war France, the main iron and steel region had been Lorraine in the northeast. But its supplies of coal and iron were dwindling and it had become a high-cost region, so that moves to the coast were needed and the companies would have had to make them.

On a second line of attack, the mechanics of Fos are also criticized. Steel manufacture failed to build up to the anticipated capacity of 7.5 million tons per annum in the 1970s (Clout 1982). The big works failed to use local labour and brought in workers from distant regions instead. Also, Fos failed to stimulate any downstream industrial development, such as heavy engineering using the steel, in shipyards, constructional steel, or car manufacturing. It was instead a "cathedral in the desert", isolated from the main areas of industrial development of the time. There were few links even to the Marseilles development as a metropole, which was based on tertiary activities and not on industry.

Generalizing the growth pole problem

The French example is inserted because it illustrates how problems might emerge, even under the optimal conditions of a country with intermediate cities that had been in some way suppressed by the growth of Paris, and would now expand, and with industries that could act as motors. In other countries, all sorts of problems exist for the operation of growth poles as planning tools for regional development.

One general problem area was in the interrelationship of city and hinterland. In a country such as Brazil, these relations might be very different, as between different parts of the country. For Brazil, the city of São Paulo has acted as a kind of growth pole to much of the eastern state of São Paulo, with a ring of industrial satellites around it involved in manufacturing for the steel and car-making industries, among others. Here there has been a positive link as dynamic industries have expanded and sought lower cost sites than those available in the city.

In the northeast of the country, the relationship is quite different. Cities such as Recife and Fortaleza are generally regarded as centres of wealth, where the surplus from traditional farming is consumed, but receiving no new investments in industry since labour costs in the city are already low and infrastructure is weak in the countryside. As a result, in the Nordeste, there are huge differences in levels of living, as well as in income levels (see Ch. 4). Apart from the industrial relationships differing, the São Paulo and Nordeste areas differ in farming systems. In the case of São Paulo, large capitalist farms exist which provide money incomes to their workers and market goods for the urban consumers. There are, in other words, good city–region economic linkages in place. In the Nordeste this is not so, and a near subsistence agriculture produces little for the urban markets, and cannot pay its workers enough to create a demand (Lopes 1977). We might summarize this general point by saying that growth poles cannot exist in isolation, and that a pre-existing positive city-to-region relationship needs to be in place in order to allow the pole to operate.

Despite all these difficulties in getting growth poles to function as originally proposed, they became part of standard thinking on regional development, not only in France but very widely in the world, and were pursued in some form up until very recently. In most cases the results have been very limited, but the policies have been attractive, presumably because they appear to be limited in need for legislation or investment on the part of government.

Growth poles could be thought of at different levels in the urban hierarchy. In France itself, the emphasis moved down from major cities to smaller towns within more rural settings. At the lowest level, small market towns might be given the growth pole status, as centres of exchange and industrial poles for their own tiny hinterland. Something of this view is contained in Rondinelli & Ruddle's (1978) advocacy of the building of market centres in underdeveloped countries. They considered the absence of such centres to be a major stumbling block for development because farmers could never move above a subsistence level without proper marketing systems, including the markets themselves and roads to link to them. There were elements of truth for such an argument in countries such as Bolivia up to the 1950s, where markets had been suppressed by the landlord–tenant contracts that required peasants to deliver all their surplus to the owner. Rondinelli (1985) did in fact use his arguments in Bolivia, along with the Phillipines, to construct a set of policies for economic development. In most

22

countries, however, market centres were only one absent feature among a large number of structural problems.

The significance of growth poles

Two main points thus came into vogue, underlying the whole development literature during the 1960s and which still play an important role today. First, the growth impulse is concentrated in single places and single sectors or industries; and secondly, there is a diffusion process taking the growth out to all other places. In relation to the economic theories, as developed through Friedmann's and Perroux's models, and to the use made of them by planners, the whole concept may be seen as an extension of neo-classical thinking. Development starting in one place moves out to other places, involving the two basic mobile factors (labour and capital) moving to correct imbalances which occur through the original impulse of development (Richardson 1973). Growth pole theory admits that the complete lack of friction on factor movements between regions is not a realistic construction, and that there are difficulties, which may be overcome through the creation of transport infrastructure and new centres in a spatial pattern away from the main centre.

It is possible to regard the poles idea as relating to Keynesian thinking, with the central aim being the stimulation of demand in regions of stagnation or decline (Chisholm 1990). This, however, overlooks the central thinking by Perroux and his followers, of establishing infrastructural conditions that put the peripheral region more on a par with the centre, and setting up major industries which will have strong linkages to other industries of the region. The productive economy, rather than consumer demand, is the focus.

However, Perroux's ideas are somewhat more complex than portrayed through the growth pole planning literature. As will be referred to later in this work, the essential idea by Perroux is not that the poles represent a good way to do regional planning, but that growth always occurs through a polarized process. Because of this, it is not balanced, but highly unbalanced, which is what provides its dynamic strength. As Higgins & Savoie comment (1988: 13), "In Perroux's highly dynamic economy, regional balance never occurs and should not occur". Thus regional balance, aimed at through growth pole theory as it evolved, should not be the aim of regional policy, and strong differences in per capita income, economic structure and growth should be expected.

For Perroux, dominant firms, sectors and individuals are bound to arise, and others will be dependent. To seek to counter this will be inefficient and will hinder development. This runs against the usual use made of his ideas, in creating balancing poles within poorer regions. Not that Perroux made much contribution to the study of geographical space, for most of his work was located in a multidimensional economic or power space. This might or might not be translated into what he termed "banal" (geographic) space. Another point is his emphasis on the firm, the *firme motrice*, usually translated as the propulsive firm, which would push development along. His emphasis is thus microeconomic, at the firm level, rather than macroeconomic, referring to the actions of governments or planning boards. To transfer from the observation of individual firms to regional planning was perhaps a major error in the use of the theory.

At the broader level, few writers have understood Perroux's message that polarization is simply a process that can be observed but not planned. One exception is Storper (1991), who reviewed the São Paulo metropolitan area and its dynamic industrial growth patterns. Decentralization of this massive conurbation is desirable, but for Storper, difficult to actually plan through setting up new centres. Instead, he was able to detect new centres coming into being, in industrial satellite cities at some distance from the city, which would eventually compete and help to decongest the city.

Perroux's ideas are of his time. In the 1940s and 1950s, France was still recovering from the war and reconstruction meant that the most dynamic industries were often those involved in this process: the iron and steel industries, petroleum refining, petrochemicals, and other heavy industries. Since transport costs were still high, these industries were located close to their hinterlands and had an immediate impact on the hinterland region. What Perroux could not anticipate was that the large firms, often nationalized industries, would come eventually to act more as giant bureaucracies, and become very inefficient. They would also, because of their bureaucratic nature, fail in the long term to produce entrepreneurs and innovators. Transport costs would also reduce so that there was no longer any direct relationship with the hinterland of these industries.

In the last 15 years, there has been a variant of the neo-classical view, based on the move from raw-materials producing (poor) countries to industrialized (and therefore rich) countries. This is the theory and policy of export promotion or export substitution. Central to this model is the view that countries must move away from being simple raw-material producers, and this is a natural evolution of the economy. But it rejects the first attempt in this direction, which was to close off countries by import substitution, starting manufacturing behind high tariff barriers that make imports artificially expensive. This works against the idea of the open market economy and against competition, with its healthy promotion of efficiency.

Instead, countries and regions must move towards industries that are for the export field, where they must compete with the exports of all other countries. In an open economy, a country or region will be able to compete successfully in industries where it has factor advantages such as a skilled labour force, and these will become its natural export strengths. This is the message of Paauw & Fei (1975), who also advocated that, rather than rely on the natural movements of the economy, a government should intervene somewhat to support the move to industrial export promotion. Their model is criticized for some opaqueness about how the transition to the new economy is to be made, and about how many people would be left behind in what appears to be a very unequal new world (Friedmann

1988), but as we shall later see, it is a working model for what has actually happened in East Asia.

Summarizing the right-wing view

For the diffusion of development or right-wing view, the agents of development are centres, most typically large cities, and from them development impulses spread out to peripheries. In these cities or centres is accumulated wealth, market structures, advanced technical know-how and information systems, entrepreneurship and management skills. They thus become responsible for the innovations of technology and for their diffusion outwards in a process of modernization. The industries of these centres offer the high wages that attract migrant labour, and selectively take the best human resources of the country.

In the long term, however, labour and skills acquire higher value in the periphery, and are attracted away from the centre to the regions. Capital too, which was accumulated at the centre in various ways in the initial stages of development, moves out to other regions because of the higher returns there. It is also important to note the aims of development under this general view. In so far as aims were described, development was to raise income or welfare, and development aid was to reduce the inequality created by the first development. In the views up to 1980, there was little presentation of the idea that aims might differ between human groups; nor was there any discussion of the idea that inequalities might be a healthy characteristic, especially if they induced movement from declining sectors or regions into more dynamic ones. Paradoxically, while the unbalancing nature of development was agreed and development fostered, a kind of ethical consideration led development aids to be concentrated on dulling the stimuli created by development inequalities.

Finally, there are the policies that result from the theories advanced. Strictly speaking, a policy of no action on the regional level could be justified, because interregional differences are seen as dissipating themselves without intervention. More generally, modest interventions to oil the wheels of industry and commerce have been seen as desirable, either to improve transport and other infrastructure (Richardson 1973), or to help the performance of existing private firms and provide a "level playing field" among a host of firms in open competition for markets.

Policies for growth poles could be seen as more interventionist, relying as they did on the insertion of specific industries at specific points in a country and their building up through the provision of infrastructure. But they still belonged to the generally positive view of capitalism as benevolent and worthy of expansion out to other regions from its first centres.

The critique of the whole diffusionist or right-wing development theory is mostly in the left-wing analyses that follow, which argue that political forces need to be brought into the equation. To this might be added that a more fundamental criticism has been the purely spatial character of the analysis, treating all places as equal and then applying economic rules to demonstrate what will happen to them. This overlooks all the other attributes of places, which are paradoxically the traditional field of geographical enquiry, and which require a more detailed analysis (Gore 1984). Such a critique is valid, although the spatial type of analysis is something that stemmed from its foundation in economics and the need to focus on certain key variables.

Radical views

Radical or left-wing interpretations of development are not directly comparable with orthodox or right-wing ones. They start with different premises, use different data and methods of analysis, and their findings neither confirm nor deny the orthodox ones. For most writers, the analysis made is directly historical, using case studies in the real world, rather than the economist's reduction to a few variables and attempts to quantify the process. Radical views do not deny the capitalist engine for growth and development. What they challenge is the identification of main factors, and instead of capital, labour and land, they insist on power, of individuals and of organizations such as governments, in controlling the course of development. To do this, they necessarily have recourse to historical and political data.

Dependency writing

In this version of the genre, the central idea is that development in the Third World is not a matter of abstract capital and labour, but of controlling forces, notably the landowners and administrators of colonial lands, and the colonizing forces of the metropolitan country. This line of thought, developed first in Latin America in the 1960s and translated and developed in English by Andre Gunder Frank (1967), became influential as an interpretation of the evolution of those countries that were colonies of European powers. For Frank, capitalism as an advance on feudalism was established at an early point in time, perhaps from the 16th century, and dominated world development from that time. For the economic historians who were the chief exponents of dependency theory, capitalism became a structure within which the whole of the modern period could be interpreted. Both development and underdevelopment could be explained by it.

As with the right-wing view, a centre–periphery structure was considered, the centre or metropolis (Frank here using a term from the Latin American literature with a different meaning to its general acceptance) being the country or region where capital was accumulated, developing industrial firms and relying on the periphery for raw materials to supply these firms. The centre became prosperous because it could monopolize the manufacturing industries while relegating the primary products to the colonies, or economic colonies.

A classic example was given by Frank (1967) concerning colonial Chile, which was a wheat producer for the urban market of Lima, Peru. The trade in wheat starts with the dependency links from an estate worker on a large wheat farm in Chile. This worker spends most of his time working on the estate for the landlord. In exchange, he is allowed to cultivate a small plot for himself. Any wheat he produces for sale is sold to the landlord, under the contract he has made, and he cannot sell to merchants who might give him a better price. Thus his dependence on the landlord is complete. The landlord in turn sells his own and his worker's wheat to a single merchant in the local town. Again, there is complete dependence because there is no effective market. The monopoly buyer (monopsonist) can set the price low because there are no alternatives. This merchant sells on, probably to another merchant in the port of Valparaiso, also a monopsonist, who arranges the transport of the wheat by ship to Lima. In Lima, an official buyer again controls the trade and acts as single buyer. The chain is completed by a second set of dependency links between Lima and metropolitan Spain. Spain decrees certain items to be tradeable from Latin America, largely raw materials, and there are official merchants designated who are responsible for the trade.

In this way, Frank portrayed a picture of unequal exchanges, in which each link lower on the chain is paid less than a fair price, and is impoverished by the higher level, so that Spain (and to some extent Lima) is enriched as the countryside and the periphery in general are impoverished. The argument operates at the level of the individual and firm, and also at the level of nations. The resulting course of interregional differences is a growing trend over time, followed by the maintenance of a high level of inequality between regions or countries.

A critique of such a view, in opposition to the neo-classical model, may be precisely that it does not conform to any model, and by using historical data, cannot be reduced to simple variables, each case study showing different situations. Other critiques relate to the problem of whether what Frank described is capitalism, or in fact historically accurate. A good case might be made that the conditions of Spain's colonial empire, like those of the other colonial powers at the time, were those of mercantilism, not capitalism: a highly controlled economy, with the mother country placing limits on the objects traded, on the partner countries allowed to trade, and on the merchants licensed to conduct the trade. Mercantilism gave way to more open systems in the nineteenth century. A standard criticism of the model is that it focused on trading relations rather than on production, and that trade alone could not be an explanation of either development or underdevelopment. For many countries, such as the thinly peopled ones of North America and Australasia, the production of goods from their vast resources was of vital importance, as well as trade. Another critique, coming from within the left-wing school to which dependency belongs, was that it failed to look at class divisions and the conflict of classes, in classical Marxist manner.

Spatial models

As with the neo-classical view, a spatial model can be constructed of the dependency interpretation of world development. Central countries, like those of western Europe in the fifteenth to nineteenth centuries, were enriched and developed as they engaged in industrialization and new technologies, their industrial march being fuelled by cheap raw materials from their colonies. Peripheral countries, according to Frank in one of his more controversial statements, actively underdevelop, or have negative development, as their pre-existing artisan industries are forced out of production by the influx of metropolitan goods and the refusal to allow the domestic industries to export.

Slater (1976) produced a simple model of the modern spatial structure of dependency, showing a centre which was linked directly and without intermediaries to each of its colonies or dependencies, explaining that the lines enabled the centre to exert control over each dependency, whether through economic forces or political power, while none of these dependencies had any means of linkage to its neighbours which might join together to form a concerted opposition to the centre.

Internal colonialism

This centre–periphery model is mostly international in level, but it may be applied also at an intra-national, regional level. The internal colonialism model of Gonzalez Casanova shows one way in which a regional dependency can be elaborated (Gonzalez Casanova 1969, Walton 1975). In rural regions of Mexico, an elite rules the local economy, white or mestizo in race, living in the towns and cities, and owning the rural land as well as the urban trading houses which receive farm products. The rural workers are either landless or live on tiny farms that they rent from the elite group, and have rental obligations, including payment of rent in kind, which ensure that they can never emerge from the depths of poverty. The conditions of dependency are comparable to those of Frank.

For a developed country, the model was elaborated in Wales (Wyn Williams 1977). He presented the Welsh area as an internal colony for England, especially South Wales, where economic development had been led by English firms and entrepreneurs in the coal and metalworking industries, and the economic surplus extracted for England, while a cultural suppression of Welsh identity was going on, including the virtual extinction of the Welsh language.

Problems in the dependency model

Difficulties in the translation of the dependency model into the twentieth century, and into a spatial frame, were debated widely, and by geographers particularly in a series of articles in *Professional Geographer* in 1982 and 1983 (e.g. Reitsma 1982, Smith 1982). One strong contribution by geographers, notably Bobek (1974) and Lacoste (1975), was the idea that the internal set of social relations within a country could often be more significant than any external dependency relation. By historical analysis, Lacoste believed it could be shown that internal elites, the controlling social groups within a country, were of great importance, and that these groups were often in existence before colonialism, so the colonial structure could not be accused of establishing them. Such groups held most of the land, the important primary resource, and prevented its better use by others. They also held the country's capital and were able to exploit the third factor of production, labour, by controlling wages and jobs. They also controlled political power and resisted any diminution of this. Any surplus or profits made in the economy were spent by them on luxury goods and imports, rather than being reinvested in the country, so checking any developmental initiative. Bobek identified the existence of a rentier group as the main social feature of many Asiatic countries, who lived in towns from the rents of land and machinery, or from loans to peasant workers. This group did not work, but lived from the work of others, while limiting its freedom. For Lacoste, the difference between development in British Commonwealth countries such as Canada and Australia, and countries such as those of Latin America, lies in the presence of elites in the latter, as opposed to egalitarian immigrant groups in the Commonwealth countries, moving into effectively empty territories. Bobek and Lacoste's views are well summarized in English by Reitsma & Kleinpenning (1985).

World systems

A more elaborate spatial model came with the evolution of dependency thinking into world systems analysis. This latter construct places modern development into the same light as the economic historians had placed colonial development, adding new forces of exploitation, notably those employed by the large firm, the multinational corporation (MNC), rather than the colonial government as with the historical cases. The MNC methods of control are those of a player in an elaborate game, using different countries for specific roles in the production process according to costs: for example, densely peopled poor countries for assembly operations which use much labour; developed countries with large pools of skilled labour for component manufacture; metropolitan countries for research and design and for administration. These companies also exploit government grants or tax remissions for locating factories, and locate often to overcome tariff barriers and penetrate otherwise difficult markets. They transfer products and materials between branches of the company at artificial prices so as to be able to declare losses in countries with high tax levels, moving production away from a country as soon as it becomes unprofitable. In world systems analysis, countries may be described as belonging to the core, semi-periphery, or periphery (Fig. 2.2),

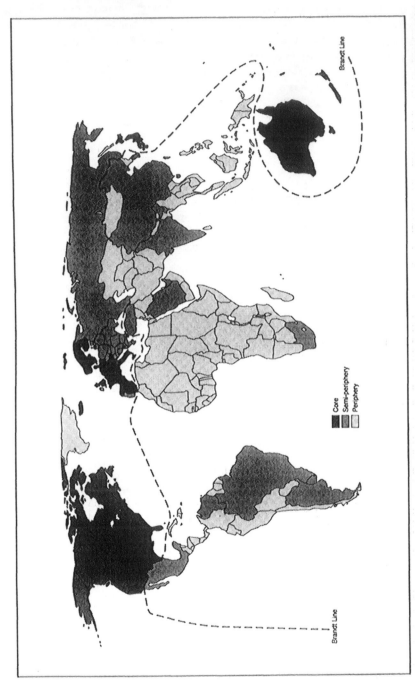

Figure 2.2 The core, semi-periphery, and periphery are mapped in terms of total and per capita GNP, which gives a rough index of global economic power. The Brandt Line makes the still simpler division into two camps, north and south, which originated with the Brandt Report of 1988.

according to a set of measures which uses not simply their level of wealth, but the degree of penetration by foreign firms, types of exports and imports, and the strength of capitalist tools such as financing through stock markets and finance houses.

For geographers, the world systems model has been most simply elaborated by Knox & Agnew (1989), drawing on an extensive literature from economists and students of political economy. It does present a descriptive framework for study of developing countries, and identifies a leading force, the international firm, which determines much of the development process.

It can also serve to direct attention to some particular side-effects of development. For example, the environment has some importance in a world systems kind of model, since it can be noted that some kinds of manufacturing process are "exported" from the advanced countries to the poor, because of their toxic effects. Waste disposal and toxic chemical processes are more easily tolerated by municipalities in the poorest countries. In addition, for processes such as deforestation (see Ch. 4) there is little local control or concern over soil erosion or the loss of biodiversity.

Global cities

World systems theory moves away from colonialism and nations, to firms. As a geographic mechanism that enables the operation of these firms, Friedmann (1986) described a special urban network, that of global cities, where the major international firms placed their offices and which thus acted as centres for regional economic activity. These were not just any large cities, but ones with particular characteristics. They were:
  (a) major finance centres;
  (b) headquarters for transnational corporations;
  (c) homes to international institutions – banks and agencies such as the World Bank and the United Nations;
  (d) strong in the sector of business services (financial services, legal, accounting, advertising, marketing consultants, and the like);
  (e) major manufacturing centres with high-level functions;
  (f) transport and communications centres;
  (g) possessed of a large population.

Friedmann was also able to identify world cities or global cities at two levels of importance. As front-line world cities he included London, Paris, Rotterdam, Frankfurt, Zurich, New York, Chicago, Los Angeles, Tokyo, São Paulo and Singapore. This list leaves out many large cities, such as the cities of India, those of northeastern Brazil, or those of Australia, or even cities like Manchester and Osaka within the centre countries, as cities that do not rank high on international function, although they have important manufacturing or other roles within their national boundaries.

31

Other writers have added further features, such as an emphasis on research and educational functions, the existence of exhibition and conference centres, and further identifiers. King (1990) makes much of the special kind of population inhabiting such cities. There is a large wealthy elite group, an international executive which creates a wealthy region in the city for itself. There are many white-collar workers, and few blue-collar workers; the growth in the city is in office space and high-class residential space, excluding or pushing out poorer groups, and creating conflict. This conflict also calls into being a large informal sector, providing cheap services and goods. The city authorities have problems in controlling such a city, with limited urban budgets. One main point about such cities is that they can act as relay points between the centre countries and the periphery. Global cities obviously exist, aside from any overarching model about the nature of the development process, and they present a problem of a special kind for the countries that host them, as well as a challenge for study, because of their special population characteristics and economy (Findlay et al. 1996).

Spatial divisions of labour

A left-wing model more relevant to the regions of Europe today is Massey's (1984) "spatial divisions of labour", which was developed for the British case. According to this model, which derives from a broader model of uneven development, a neo-Marxist alternative to Frank's dependency theory (Weaver 1984), in the early industrial period of the nineteenth century manufacturing industry was organized by regions, with a complete structure in each for the lead sector. Thus, for example, southern Lancashire became a cotton textile district, with a complete set of vertically integrated industries, from cotton-washing, carding and spinning, through to weaving and finishing industries, often in different mills with different owners. But in the present century this arrangement has been changed so that regions become specialized, not according to sector but according to stage in the general industrial process. Thus London and the southeast of England have become the main industrial centre for research and design, and for company headquarters, especially of the light industries that have emerged in recent decades such as electrical and electronic goods. The old industrial regions of the nineteenth century have tended to become regions of component manufacture, using the large pool of labour skills in these regions. Some regions, notably in the outer periphery of the north of England, Scotland and Northern Ireland, have become largely assembly regions, their skills having declined over long periods of depression, and their wage costs now being lower than elsewhere.

This model adds some complexity to the overall picture, with the assertion, also presented in the world systems writings, that the role of any specific region may vary over time: that for one period and one technology it may be central or nearly so, while at another time it may be peripheral. This means that each region has an individual character and set of problems. Thus South Wales, at one time a major coal-mining region with related heavy engineering, became a problem region from the 1930s with the decline of coal, but in the 1980s has achieved a new position with much light industry, since this region lies as an appendix to the Bristol region at one end of the M4 motorway high-technology corridor. Centre and periphery are defined, as with the world systems or the NIDL (new international division of labour) models, in terms of the role each can play in the vertically integrated production system.

Nor can the position of a region with respect to an industry be predicted. Massey (1984) used the example of the shoe industry to show the fickle (from the point of view of specific regions) changes in location. In 1800 the industry was located in London. It moved out for cheaper labour to Northampton among other places. Industrial organization was into factories in the town, cutting the leather, and domestic work at home by women, often in villages, stitching the shoes together. This changed later to bring all the industry into the factories. In recent decades, the industry has decentralized out to the smaller towns and villages.

Points of comparison between left and right

In the various left-wing analyses, the idea of declining inequalities through some process of balancing factor movements is denied, which clearly differentiates the whole school from that of the right wing. There are, for example, critiques of the growth poles strategies from a strong left wing point of view, to show that these are set up with the needs of large-scale capitalism rather than regions in mind, and that for the poor countries, the whole concept of growth poles is inadequate or inappropriate (Conroy 1973). For world systems, there are possibilities for any individual region or country to improve its lot, but the system remains in place, with a powerful centre that controls or restricts the development of the periphery.

There are some points of agreement between left and right, however, which should not be overlooked. Most importantly, the analysis from both sides is that development is an exogenous process, relying on some distant source or centre. Where they differ is in the interpretation of what the influence of this process may be. For the right, it is a beneficial influence, bringing forward development through the natural movement of the factors of production. In the long term, the higher production and consumption levels of the centre are relayed out to the peripheries. For the left, the alternative view is that the effects are negative: that the growth of the centre attracts out both the capital and the human skills of the periphery, which end up in the centre. Because of the control systems exercised in the centre, the development process restricts any progress of the periphery, and it remains an area of raw-material production and low-skill manufacture.

With regard to the aims and the policies that emerge from these analyses, for the left wing there is a strong view that inequality between groups should be eliminated, enunciated more clearly than from the right wing. Much of the policy to enact the levelling process is sectoral, not regional, such as fiscal policy to level incomes through taxes on wealth or possessions, or welfare provision of schooling, medical aid and public transport, to ensure access to services for all at modest cost. Policy is thus towards improving consumption for the poorest, rather than worrying about production. Some policies within this framework came to have the tag "demand stimulation", following the ideas of John Maynard Keynes, since increasing the consumption of a broad sweep of the population would in turn increase production levels. Concern for welfare links neatly to demand stimulation policies. But regional policies have also been employed, as for example in most West European countries over the 1960s and 1970s, followed in the late 1980s and 1990s by a more forceful EC policy set for the regions. EC policies have also identified some other regional problems beyond those of non-convergence between the regions, such as the special set of problems for old industrial regions which need to convert to new industries.

It must be acknowledged, however, that some versions of left-wing theory do not admit of any easy regional policy set. World systems theory sees the penetration of international capital into all corners of the modern world, and any country or region wishing to improve its own lot must break out of the system, a difficult thing in itself requiring strong barriers to trade, and of dubious results given the experience of those countries, like Tanzania and Cuba, that have tried to isolate themselves from outside (Western) influence in recent decades. For regions to accomplish an even partial closure, as suggested specifically for Development from Below, seems an even more daunting and perilous task.

On the right, the implication of the analysis of gradual balancing through the influence of the market does allow a lower level of regional policy and planning. However, two kinds of policy are allowable within this analysis. First, a policy set that improves infrastructure, and reduces the costs of distance and separation, is in line with an analysis that recognizes real costs and transport barriers to overcome. Thus roads and communications improvement, perhaps with growth poles to spread development, is a part of such policy. Secondly, policies to overcome specific problems through changing circumstances are relevant. Thus, the policies of the EC for old industrial areas in transition, for the reform of farming areas dominated by small farms, or for areas with specific environmental problems, are relevant.

Resources and development

Both left- and right-wing views do have in common their focus on the mobile factors of production, labour and capital. Another vein of development thinking with special relevance to the study of regions has been the ideas about development based on local resources. These ideas were given a more formal grounding, as "staple theory", in the work of some economic historians, notably

Innis (1930) and North (1961), concerning the "export base" of nations and regions. The idea or theory was that a region, especially a newly settled region, would develop and become prosperous as and when it discovered and was able to export its physical resources. The outstanding growth of the US Pacific Northwest, for example, was a function of the export of, first, fur, then timber products, then wheat, and finally the enormous electrical energy potential from its rivers, in the form of aluminium and the aeroplanes made from it. Each export staple brought new money into the region and enabled service industries to grow and feed off the staple. It was never fully possible to justify the theory, only to indicate particular regions where it seemed most to apply. At the theoretical level, it was found that exports from a region depended heavily on the size of that region; the smaller it was, the greater was the importance of exports. For the same reason today at national level, it is impossible to compare the exports of, say, Singapore, with those of the USA. The latter trades its goods and services mostly within the boundaries of the state, whereas the former must always depend on exports to the rest of the world.

There are few regions where such a theory is applicable today, where undiscovered or undeveloped resources are to be tapped. Chisholm (1980, 1982) did show that resources might at times have critical value, but that the value changed with the economic cycles, so that periods of stagnation in manufacturing, and low value for manufactured goods, were accompanied by high values for resources, and vice versa. High resource prices might put regions rich in resources, notably today oil, gas and some metal mineral resources, at an advantage, but they might then lose that advantage with the change of cycle to revalue manufacture.

For the countries of East Asia, and for their comparison with those of Latin America, it will in fact be necessary in later chapters to stand the resources argument on its head. Those countries of Latin America with large resources of minerals and open land, such as Argentina and Venezuela, have had a poor record, while countries in the Far East with very limited resources, like Singapore and Korea, have achieved rapid economic growth. To some extent this has to do with the pressure that resource poverty imposes on human effort and ingenuity. Development comes from need, historically, as is argued by Wilkinson (1973), and it probably still does today. Resource abundance also provides too much of a temptation for politicians and governments, who are liable to be corrupted by the power and wealth given them through sale of the resources. This argument is particularly valid for Latin America.

Anti-development

Since the collapse of the Soviet Union in 1989, there has been a crisis in development thinking, particularly for the left-wing view. Shuurman (1993)

35

writes of an "impasse in development theories", following the powerful criticisms of Marxist as well as modernization theories through the 1970s and 1980s. The 1990s crisis has come about not solely because of the failure of communism in one country, but also because of deep concerns over the nature of development itself, the strong challenge and apparent success of liberal policies, and the rise of post-developmental or anti-developmental arguments. We may explore briefly the last of these elements here.

In the 1990s, there is a rising tide of criticism about development which argues that the main thesis of development, largely from "us" to "them", is itself merely a thesis, a set of ideas, rather than a real process. Escobar (1995), for example, writes of the "invention of development", implying that the concept has been produced out of the minds of politicians or academics, rather than out of an actual process existing either now or in history. In this book we will assume the reality of the developmental process and the objective value of efforts now being made. However, it is worth examining the opposing case which has been made over the last ten years or so.

Escobar's view, to cite one foremost writer in this idiom, is that there is a "discourse of development", a set of ideas and propositions which are accepted and which link logically together. This discourse starts with the proposition that the development problem was discovered, perhaps as late as the end of the Second World War, and stated as being the problem of poverty, linked to such processes as population growth, the small farm size of the poor countries, and the problem of the landless. This then is seen as a technical problem, which may be overcome with technical means requiring special knowledge. In turn, this leads on to the statement that the developed countries have the necessary skills and knowledge in order to be able to solve the poverty problem, in their public agencies and universities, together with the capital that is to be used in the development process. This leads on finally to the need for the developed countries to intervene in the poor countries in order to sponsor their development. There are other versions of the discourse. Some writers have traced the development concept back to the 1920s or 1930s; the point is that the concept is relatively new and in no way innate to human society. Emergence in the 1930s may be presented as an exercise in legitimation. For those countries such as Britain with a large empire, or the USA with an informal empire, development needs provide a legitimation of their continued intervention in their colonies or in countries where they held strong influence. For Slater (1995), the discourse of importance is that of modernity, which begins with the Enlightenment, the period in the eighteenth century when restrictions on rational and scientific writings and research were generally lifted, and when many writers discussed the possibilities of a continuous advance of humanity. Development is thus seen as part of this movement towards modernity.

The idea that such a discourse should exist, independently of the facts, comes from a number of French linguistic philosophy writers of the 1960s and 1970s (Peet & Watts 1993). For them there is no necessary correspondence between discourse and objective truth. The challenge to modernity comes from a school of thought that has taken the period from the Enlightenment to the present as one of modernity or modernization, and thus places itself as the school of post-modernist writers. A broader view is taken by Cowen & Shenton (1996), in looking at development in many different countries, and the statements of politicians and others. Rather than during the Enlightenment, they place the idea's beginnings in nineteenth century industrialization and the wide variety of problems in adapting to industry in different countries, including many colonies. These writers broaden the scope of enquiry about the origins, although they still wish to find a formal and literary source, rather than a concern which is innate in human interests.

Rather than an invention or a set of ideas with limited time relevance, development for our purposes may be taken as an idea with continuous relevance to societies throughout history. While it is accepted that the post-war era has seen some special arguments about development, some of them deliberately skewed in order to favour powerful nations in their attempts to legitimize their own intervention in poor countries, there is an objective process which is observable in many countries, irrespective of whether there is any intervention by others. In part this process is technically led. As machines have been adopted, as huge resources of power and raw materials have been tamed for humanity, and especially as great increases in communications have taken place, more and more people have become aware of societies beyond their direct cognizance, and this has led to an awakening of interest in their own improvement. This awakening is a driving force for development that it is hard to quench.

If we take a long view, there are also many counteroffensives against some of the effects of development, so that it is possible to see the post-modernist view as simply the latest attack on the development process. In the period since the first Industrial Revolution, the protests have been more vociferous, and with good reason in many cases. Robert Owen, for example, a manager for modern development of textiles in Scotland, eventually rejected the damage done by the social reorganization for factory work in Lanarkshire, and attempted to set up new social structures on the Mid West frontier in the USA. Later in the century, William Morris argued for a more holistic approach to life than was permitted by the factory system of his time, and returned to pre-industrial models of craft industry to put his ideas into practice. In the present century, the ideas of Development from Below, or of "another development" (Hettne 1990), are more sophisticated and more theoretically based ideas in the same vein. The most positively oriented ideas come from writers who seek to help specific groups such as rural dwellers, women, ethnic minorities, and the environment as a special case of a minority; these have considerable relevance in checking the market-related mainline theories. On the other hand, it is possible to detect in much of the anti-development lobby a concern for the minority interest that actually overlooks the main thrust of development. The focus of such writers as Escobar (1995) in his general review, or detailed work such as that of Routledge (1993,1995), picks out rural disadvantaged groups and privileges their position. At the present time, it may be argued, the centre of development is not in the rural areas nor in minority groups. There is instead an urban and industrial focus. In this book, development is also argued to be a highly concentrated affair, so that peripheries are always liable to appear. Complete equity is an unlikely state of affairs.

Conclusions

Summarizing some of the above, it would appear that the different ideas presented are for the most part not directly in conflict with one another, but concerned with different subjects. On such matters as the origins of developmental impulses in urban-industrial centres, most writers would agree, as they would on the advantages presented by free movement of the factors of production, people and capital resources. A central point of agreement, tacit or formally stated, is on the exogenous nature of development for most of humanity; development or its lack is derived from somewhere outside the core, centre or mother country; relatively little credence is given to endogenous, home-grown development, which is perhaps harder to study, isolate or plan for. In later parts of this work, this stress on the exogenous will be called into question. Some of the fundamental critique emanating from the anti-development writers just discussed can be assimilated into the earlier case made for Development from Below. Anti-development is very often merely against development organized from elsewhere.

In terms of content, although there are large differences apparent, neither side pays much heed to one of the main present-day concerns, the environment. Neoclassical theory leaves land out of consideration. The export base concept developed for regions did use the idea of a primary resource as the engine of development, but it was an obviously flawed concept because of its very simplicity and its limited applicability. Radical theories, on the other hand, only mention resources as a possible target for central powers that seek to control them. A recent past with wide availability of natural resources from varied sources has perhaps made them appear unimportant for development. But a growing concern with pollution means that negative environments have their own cost, and cyclical materials shortages may force more attention to the positive aspects of environmental use.

Nor has there been much attention given to technology by most of the writers. Traditional attention has been focused heavily on the use of capital and labour, which overlooks the huge contribution made by changing technology to development. Yet whole theories of development have been constructed on the basis of technology change rather than change in capital and labour.

Concerning the aims, there is a general consensus that inequality between regions is undesirable. Right-wing views are generally only that convergence, the righting of inequality, is a good thing, and that it may need some help through, for example, transport and communications; in the long term, the market will achieve convergence anyway. Left-wing views are for more powerful action towards convergence, either by moderate governments providing extensive aid to poorer groups of society and poorer regions, or by cutting off nations and regions from outside control. Neither side gives much attention to the possibility that inequality is a natural condition and that it may be quite healthy in promoting a dynamic process of growth. This possibility will be examined again at various points. As for policies, the level of agreement is quite high. There seems to have been a dominance by technical or professional views as to the kinds of policies possible, rather than political views, so that left and right wings have employed many of the same policies over the years. At present, regional policies that support innovation and the generation of small firms are welcomed by both sides. Policies on both sides in the 1950–80 period emphasized manufacturing industry, in a blanket acceptance of any kind of industry. Since then, much more emphasis has been placed on the tertiary and quaternary sectors as the central dynamos of development in many of the advanced countries.

CHAPTER 3
Industries and firms

From the previous chapters is taken the idea of concentration as something that has happened widely in the process of development, and which has a positive value in this process, as observed in the newly industrialized and developed countries. Other ideas are those of the appropriate and changing industrial mix, the limiting or helping role of the state, and the varying fate of regions in the process. Such ideas are, however, of a process viewed in general, at a macroeconomic or national scale, without any examination of the role of individuals, firms or industries in it. It is necessary to examine the microeconomic level, that of individual firms, because they are the basic units of production. Some of them are in any case very large, and their influence may be more than that of countries; but the collective functioning of a large assembly of small firms is also of interest and this theme will be developed in this chapter. Economic mechanisms operate both at the firm level, and at that of groups of firms and the institutions that link them. Some other structures are found at a second broad level, that of whole industries, such as car-making or steel.

In what follows, models of the spatial structuring of development at the level of products, of firms, and groups of firms or whole industries, are introduced. None of these models has a close fit to reality in any country, but each contributes to the analysis of changing industrial structure.

Differentiation of production between centre and periphery

One way of viewing the changes in economic organization in recent times is in terms of a "new international division of labour". The term comes from work by German writers (Fröbel et al. 1980), published originally in German in 1977. Their research started from the alarming finding, for a West Germany that had seen constant growth since the Second World War, that some industries, notably textiles and clothing, were in decline in Germany, and that the whole textile industry was being effectively "exported" to Third World countries, because of their lower labour costs, and lower labour rigidities because of less unionization. Other industries were retained in the advanced countries, but the export of jobs was causing problems in the heartlands of manufacturing industry. Some firms

41

exported the low technology, labour-using parts of their industrial process to poor countries, and kept the more sophisticated parts in the home country.

This was not really a new finding; the dependency writers had for some time been basing their critique of capitalism on the unequal exchange between a centre with the manufacturing, and a periphery condemned to producing primary products. This was the older or classic division of labour, separating primary producers from manufacturers. The change observed by 1980, to the "new international division of labour", was the move of all countries towards more technologically demanding activities. This meant that while there was continuing technological change and new industries in the centre, some less demanding industries or processes, and no longer just primary production, could be located in the periphery. If, in the past, Germany made steel and Brazil produced the iron ore for it, now Brazil might also make basic steel ingots, and Germany would work these into engineering products such as cars.

The rules of comparative advantage still apply, so that places with an abundance of labour are more likely to have labour-using production, and those with more capital and high levels of education and training (human capital) are likely to retain the technically advanced industries and processes. Textiles and clothing production are a classic case, as the technology and capital requirements are relatively low, and labour costs quite high, so that this industry tends to be exported to developing countries. Often the export is within one firm, which establishes a branch plant in a foreign country.

The same division may be arrived at by looking at individual products or groups of products. Vernon (1966) described the "product cycle" to show how changes in production would occur over time, with a product that starts with an innovation, grows into many experimental forms, then becomes standardized, and finally goes into decline through lack of any change, plus the advent of alternative products which take away its market. Malecki (1991: 126–8) described several of the cycles that occur in similar form, and some of these may be put together to show a general set of changes over time-space.

To put this in concrete terms, we may group all of these aspects together in

Table 3.1 The product cycle.

Aspects	Stage I	Stage II	Stage III	Stage IV
Technology	Innovation and experiment	Growth	Maturity and standard product	Decline
Profits	No profit	Growing profitability	Stability of profits	Decline with monopoly profit
Innovation	Product innovation	Product innovation	Process innovation	Process innovation
Geography	Home region	Home country + DC branches	Home + LDC branches	Home/DC/LDC hierarchy

terms of a particular product, like the motor car, which will also illustrate the weakness of the model. This innovation begins its stage I with experimentation, in Germany, Britain and France, in the 1880s and 1890s. New products were invented and tried on a small scale, with no profit-making at this stage. Many of the early inventors, who combined with or had already been carriage-makers and makers of bicycles and other engineering products, failed to make satisfactory machines or were bankrupted before they could become established in the market.

In stage II, from around 1900, firms began to grow and the early car manufacturing industry became established in a few European countries. Some of this early spread was that of ideas and processes to new firms, while some was expansion by the very earliest firms. Daimler, for example, the German motor company, set up in Austria in 1902 and sold patent rights to the British Daimler company in 1890, licensing Panhard to produce in France from 1891, and in New York licensing another Daimler Company from 1888 (Maxcy 1981).

In the later part of this stage, in the 1910s and 1920s, the USA became a major producer, effectively creating a second centre for expansion with two major firms, Ford and the companies that would link to form General Motors. From this new base, firms set up branch plants in Britain and elsewhere in Europe – the first Ford and General Motor plants in Europe are from this time. In the three years prior to the First World War, the USA already produced 78 per cent of world car output, and in 1929, 84 per cent of a peak level of world production. Apart from expansion, this stage saw the emergence of a differentiated industry, as Ford and General Motors became the largest mass producers and multinational enterprises, while some of the European manufacturers became small-scale producers of quality cars as a definite policy.

Stage III, after the Second World War, involved disintegration of the formerly spatially and vertically integrated industry. Production methods and components became increasingly standardized, allowing lower cost production for the mass market and its dominance by a few, up to 20, large firms. This standardization also allowed labour-intensive operations, notably assembly, to move out to LDCs because of the lower costs; as competition bites hard and profits go down, this is a cost-reducing strategy. Movement out was also to overcome national tariff barriers by producing what were ostensibly locally made cars, although this is not a central part of the model.

In stage IV, cars are made in many countries, and all sections of the industry are widely distributed. The pace of innovation slows, although in the car industry, contrary to Vernon's standard model, there are continuous changes in process as well as product which make the car a more lasting product than some less complicated ones. Continuous innovation around a complex product means that some centres of the industry have remained so for a long time. Apart from spreading out to most of the world as a set of independent producers, there is also some attempt at globalization, i.e. the manufacture of a world car or global car from components that are the special domain of individual countries or regions. This was a particular direction taken by Ford in the 1980s.

In the last two decades, a fifth stage may be added to these four. With the partial failure of the global car idea of Ford, there has been reconcentration of production for some countries and firms, following the Japanese system of "just in time". This relies on close contacts between manufacturers of cars and their components, small inventories, rapid communication of technical information, and close control over quality, all of these elements needing a physical location in proximity to other producers. This system reverses the geographical trends of all the previous stages, and would seem to question the whole product cycle idea.

Putting the product cycle into practice

The last statement, about the continuity of a single centre or central region, illustrates a general point to be made about the product cycle. Although it describes the evolution of production tendencies for individual standard items, for many more complex consumer goods like cars or computers, there are many individual elements which can each change in technology and process of production, at different rates. It is at the major centres of technological advance, once they have become established, that the changes tend to occur. For some capital goods such as fertilizer manufacture, the pace of technological change is slower, and investments in production equipment are made for longer periods of time without change. In this case, we should expect the product cycle to work. Product cycles operate most obviously for individual products such as electrical generators for cars, first made only in the central countries, then moved out to many LDCs, and finally replaced in modern cars from the early 1960s onwards by alternators. In a large industry sector such as textiles, there are different evolutions for different products (van Geenhuizen & van der Knaap 1994). In Holland, in the face of rising costs of production, the knitted goods industry has moved out to lower labour-cost countries overseas. The carpet industry adopted a strategy of technological innovation to produce goods for which there was a special scarcity value; and in weaving and finishing industries, there has been a strategy of acquisitions by the biggest firms, and a partial adoption of flexible production (see below).

If we represent the four stages of Table 3.1 as a production curve over time (Fig. 3.1), it is possible to visualize various ways in which the simple growth and decline pattern can be bypassed, even for a single item. A first possibility is that the whole cycle can be curtailed by a substitute product coming in and replacing the original. Thus, in internal combustion engines, alternators suddenly replaced generators of electricity from 1960. Another possibility is a revival through finding new applications for the original product. In textiles, the underfelt for carpets finds new applications as a mulch around garden plants, and sites subject to erosion are covered with plastic mesh to retain soil and vegetation. There are a variety of other geotextiles coming into use. Other possibilities are changes in technology or fashion which allow a product to retain, or to regain, its value. Cast-

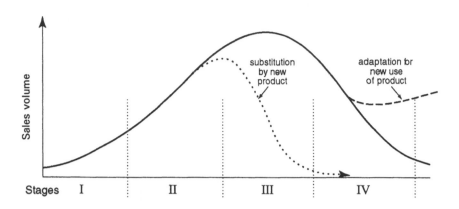

Figure 3.1 Product cycle variations.

iron stoves, popular in Scandinavia and formerly in much of Europe, have attained a new importance through a combination of changing fashion and the invention of the chain saw, making more small-diameter wood available to farmers and home-owners. What are the results of this set of processes for specific regions or countries? Innovation in the first stage causes no divergence of wealth or income, so that it presents no threat to anyone. At the other end, spreading out production to all areas again spells no problem to specific countries or regions. Attention thus focuses on the intermediate stages, when the products are being brought in at the centre, but have not, or have only partially, been distributed to the periphery. In the comments that follow, a link is made directly from product to firm to region. In reality, of course, the effects are tempered by diversification within the firm and within the region. A gross simplification is thus made to present the case.

For those at the centre, the problems depend on the organization of production, and on power systems. The theory of dependency, discussed earlier, is relevant to the consideration of power and control. Firms, like governments, may restrict knowledge of key processes to their home-workers and factories, leading to the current demands from most LDC governments that there be a technology transfer with any major new investment. But even without a power system that causes imbalance between regions and states, there may be special problems. In the case of a region that produces only a single product, with no variations and no innovation, there is an obvious danger in the product cycle because after the early stages the role of the centre is lost. Thus a region that specializes in the manufacture of basic steel is liable to become a crisis region when the steelmaking process is learnt by a variety of other countries, and the first region is unable to change its industrial structure. In this particular industry there has in fact been some innovation, such as the introduction of direct reduction techniques in the

45

1960s, avoiding the traditional two-stage blast furnace–steel furnace process. But the pace of innovation has been slight compared with the learning ability of LDCs, which now threaten all the old steel regions of Europe, such as South Wales or the Basque country in Spain.

From the above commentary, it is apparent that the product cycle poses some problems for central regions. But in reality, for most kinds of products and most kinds of regional division of labour, it is the peripheral regions that are at greatest risk from the cycle processes. In these outer regions, the labour force suffers from the lack of any attempt on the part of the controlling firm to build new skills. Since standard low-skill work is all that is required, the labour force is poorly paid, which gives it little chance to improve its lot; there is no role for, or creation of, innovators and entrepreneurs who might be able to develop their own technology and thus new firms. Perhaps the main drawback is the dependency relationship of peripheral factory to central factory, or peripheral firm to central firm. The peripheral unit has no connections to other firms or outlets for its products, and is dominated by the central firm or factory. It is often also a unit producing a single standard product, using a single type of process. This technological rigidity means that the peripheral unit, and the whole region associated with it, has no obvious means of developing the skills or the machinery to produce other goods. It is such regions that feel the blight of technology moving on. For Massey (1984), they are the victims of mobile capitalism; in reality, they are the victims of moving technology, of technological change. Rigidity in operation, whether of process or product or both, means that any change in fashion (in the case of such sectors as textiles and clothing), or change in raw materials (substitution of an expensive metal for a cheaper or a more durable one in car manufacture), may mean that the factory cannot continue production at all. At the simplest level, the product cycle leads to all countries or regions having the technology and other factors to make the product (such as the generator); in reality, there is constant innovation, which favours the central regions as a whole, especially if they have control of complex manufactures such as motor vehicles, and a choice of processes and products to put into the finished machine.

Flexibility

The product cycle is thus linked to the division of labour, and both are linked to the industrial fate of regions and countries. But on its own, the product cycle does not tell us the complete story of the structure of modern industry. In the situation for industries of relatively advanced technology such as car-making, and because there is a constant stream of new products through innovation, a semi-permanent division arises, between the tasks undertaken by productive units in the centre, intermediate and peripheral areas. Central units conduct administration, research and development tasks. Those in intermediate locations make components or even

46

finished products, but without the higher administrative and research functions. The outer regions are given the tasks of assembly of finished products. In the case of simple products, the centre may cease to be involved in the product any more and be engaged in researching new products to replace those currently being made in the periphery. Product cycles only capture individual elements in this chain of events.

The "new international division of labour" (NIDL) does not offer a much more complete idea of the spread of industry. Since its prediction is simply the movement of manufacturing out from the central countries to the LDCs, through the lower labour costs involved, this schema does not account for any special concentrations of industry, or the rise of successful industrial bases in intermediate countries or those that do not offer simply low-cost labour. Nor does the NIDL account for the rebirth or the survival of industries in the 'old' industrial countries. In Germany itself, where the NIDL spectre was first raised, statistics today show a fairly even balance in the value of textile imports against exports, and this is the same as in 1975, the period for which the NIDL model was created. Clothing is imported in a ratio of 3:1 over exports, the same as in 1975. Evidently, these industries have survived against the predictions of NIDL, and for most West German industries this has been because of their improved flexibility. Nor do the import–export ratios imply a static industry. Output value has risen from an index value of 100 in 1975, to 235 in 1989 (van Geenhuizen & van der Knaap 1994).

Today trends can be observed for some kinds of manufacturing production to be organized in more flexible ways, by firms or groups of firms, and this may be the basis for a new geography of industrial production. Many names are given to the different forms, but two broad types may be identified here: flexible production and flexible specialization. Both of these allow smaller volumes of production to be handled economically, enabling rapid changes in matters such as colour, design details, or even the processes used in manufacture. Some of these trends may be attributable to a post-modern urge for differentiation of each person's tastes from those of their neighbour. Fashion is certainly a dictator of production, although this may be a feature of increasing affluence and ability to make choices, rather than any more deep-seated changes in our world view. As far as the actual organization of production is concerned, the whole movement is regarded as one away from "Fordism" (the use of mass production lines, the excessive division of labour, the manufacture of highly standardized products: all principles endorsed by Henry Ford in developing the North American car industry), and towards something called post-Fordism, which reverses all these trends.

Flexible production

Taking the first type of flexibility, flexible production, we may outline several of its distinctive features. It may be regarded as the "high-tech" version of flexibility.

47

It involves the use of automation in order to be able to programme designs and rapid changes of design (computer-aided design or CAD). The machines used in manufacture may be computer-controlled (computer-aided manufacture or CAM), and are in consequence themselves flexible, their tasks being programmable by an operator. In the simpler versions of CAD, this involves automation for individual machines. In the most advanced examples, there is a wholly automated system of manufacture, or what has been called a "flexible manufacturing system".

Firms engaged in this kind of production move from the search for economies of scale (i.e. those obtained from very large runs of a single item) to economies of scope, obtained from producing several different products from the same machinery, switching easily between them according to demand. Some of these economies of scope may be obtained through combining with different outside firms in a network.

One well-known form of flexible production is the system entitled "just in time", in which the handling of supplies of components is of essential importance. Components are not stocked at the factory, but brought in from nearby factories on demand to meet the requirements of production for the next few hours. This system is typified by Toyota in their car manufacturing plant at Toyota City, on the outskirts of Nagoya in Japan. Here the central assembly plant is surrounded by a halo of component manufacturers, largely run by independent companies, although their main contract is with Toyota and they work in a *keiretsu*, a strongly linked set of companies which work on a more cooperative than competitive basis (Arnold & Bernard 1989). Toyota requires these suppliers to send regular small deliveries of components to the main plant, in some cases several times a day, and the assembly lines depend on a continuous inflow of components.

The system has the advantage of little waste, since faulty products may be identified at once and their supply stopped. It also provides for easy quality control as all components can be tracked immediately to their source. Further advantages are the flexibility of output, in types and quantities, as a rapid switch can be made between two different products used in different models. A fundamental advantage is, of course, the very low inventory costs. Such a system clearly depends on a closely linked network of factories with good transport and communications between them. At Toyota, lorries arrive through the day with small deliveries of a host of items, and the flows are subject to continual modifications.

Such a system has a definite spatial form, with subsidiaries and support activities constituting a kind of industrial district. This is still, however, a centre and periphery structure in miniature, the central role held by the main assembly firm. It is also true that most of the higher order functions, such as research into design and technological matters, as well as the general planning decisions such as the move to new markets, are taken by the central firm. From the examples of Japanese motor manufacturers, it is not apparent that this kind of organization of industry will lead to any new geographical pattern. Old centres can still establish themselves, and the effect may indeed be reconcentration of work in and around the central firm, rather than the NIDL pattern of moving simple functions out to peripheral regions. It has, however, been noted that in recent decades the Japanese model has been transferred to other countries. Within the USA, for example, Japanese companies have occupied a new industrial space, marginal to the old Mid-Western industrial belt, and reaching down from Ohio into Kentucky and Tennessee (Fig. 3.2). The system thus encourages some regional shift of activity. Choice of new places for manufacture apparently responds to a need to escape regions where labour unions and regulations make it difficult to establish a new work pattern and work ethic. In the American Mid West, the new Japanese plants have attracted their own Japanese component suppliers, but these are close enough to be able to supply also the older native US motor firms. Industrial relocation is thus through a gradual shift.

It should be noticed in any case that the Japanese location habits in the USA are comparable to those being adopted by domestic firms in that country (Rubenstein 1986). For General Motors, Ford and Chrysler, the three largest home companies, it has been seen as advantageous for at least two decades to move into states adjacent to the original car-making states, into an "outer Mid West" of states like Kentucky, Missouri and Tennessee. This pattern of "dispersed concentration" is thus becoming an important feature of the total North American structure. The reason for the dispersion was initially a need to supply regional markets from closer factories. In more recent years, the move is into states with more flexible labour and less union control. Similar trends are to be seen in the Japanese plants in Britain, located at new sites like Derby, Washington in County Durham, and Swindon, away from the old industrial areas but accessible to the overall industry.

The long-term significance of the Japanese "just in time" type of organization, both spatially and technologically, is unknown. In terms of spatial concentration, dispersed concentration seems to be the trend, and this is something replacing a previous trend towards a centre–periphery model, with global manufacturing involving many countries specializing in the production of components which could be put together in a number of final assembly plants. But the Toyota pattern is not being followed by all competitors. It should be noted that a leading Japanese manufacturer, the Honda company, has spread its factories widely in Japan, and relies on a more complex network of linkages between factories, which does not correspond to a cluster around a central assembler or main company (Mair 1995). Honda has main plants in Sayama, Wako, Suzuka, Hamamatsu, and Kumamoto on the island of Kyushu (Fig. 3.3). Its parts are supplied mostly from the Tokyo area, where Nissan is centred, and from Nagoya, the Toyota centre, and it uses the component manufacturers linked into these other companies. In terms of company structure, there is no *keiretsu*, the typical Japanese arrangement with a closely knit set of companies around the central one. Honda can nevertheless claim to be at least as innovative and successful as the others.

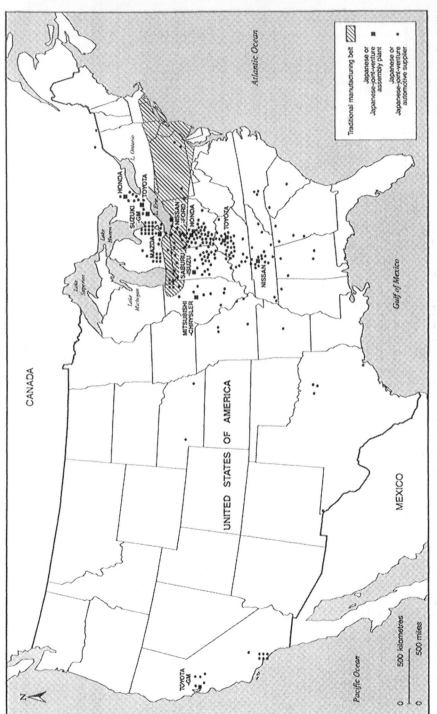

Figure 3.2 The new Mid-Western industrial space.

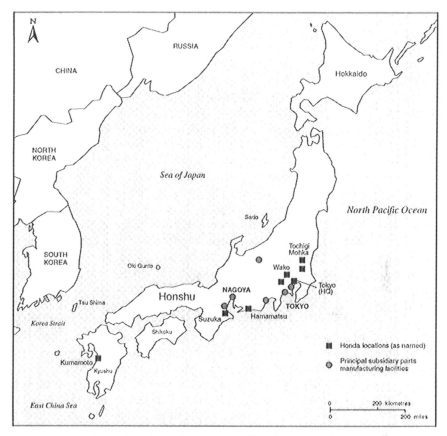

Figure 3.3   The distribution of Honda plants in Japan.

Flexible specialization

A central aspect of the flexibility just described is that it links vertically, between first-level producers of components, and second-level assemblers, sometimes with an intermediate level between. Another kind of flexibility is that derived from a more complex, more horizontal interfirm network of linkages, known as flexible specialization. The arguments at its base, some of which are also applicable to the previous kind of flexibility, are best made by Piore & Sabel (1984) in their wide-ranging review of the economic history of manufacturing in Europe.

They start from a basic critique of capitalist development similar to that of Marx. Capitalism leads to recurring crises of production because of the growing antagonism between capital and labour, which cannot be solved by negotiation. Economies of scale, coming into effect with some force in the late nineteenth century, constantly encouraged bigger firms and units for capitalist production, replacing an older organization of workshop industries in industrial districts. Big firms and factories spawned big labour, in the form of trade unions seeking to improve their members' conditions of work. Over time, the conflict between labour and capital grew over the structure of production, the allocation of jobs, the rights of workers and their pay. This became more insistent as the actual jobs became less meaningful, with less skilled work and less responsibility for the individual workers. Crises were thus produced partly by labour. They were also produced by capital, which tended to produce ever-larger quantities of standardized goods, and was unable to reduce or to change materially the kind of production. This led to crises of overproduction when tastes moved away or were reduced. Piore & Sabel saw a way out of this problem, not in the Marxist radical solution of revolution to overturn capitalism altogether, but through the adaptation of the older nineteenth century organization of production, under principles of flexible specialization. In this older system, industrial districts were made up of many firms in the same general sector of industry, such as textiles or light engineering. Each firm had a degree of specialization, but used simple machinery to turn out batches, and was able to be converted to a different product at relatively short notice. Further flexibility in such a system was endowed by moving the work amongst the different specialist firms. In textiles, for example, a different kind of dye process was needed when the material was changed from heavy to light grade wool. This could be met by using a different firm specializing in the process. If a normal run of business was disrupted by a sudden demand for a large quantity, flexibility in cooperation allowed the contracting firm to share the order with several firms in the same business, or to subcontract some of the business to outside firms. External economies were also available in the industrial district, which might have specialist colleges offering relevant courses to train workers, and offices to act as an employment exchange.

The social context

Industrial districts were also normally characterized by a degree of social support to the economic system, with a variety of organizations which could help firms to keep in contact with each other, such as trade organizations and chambers of commerce; or employee organizations, providing insurance and help to the unemployed, or finding job opportunities. Beyond the direct organizations of business, a broader measure of support might be available through organizations such as the Catholic Church, cited in the case of the Italian examples of flexible specialization. Another source of social support is the great urban municipalities, which in the Italian cities, and elsewhere in Europe, made special efforts to support their chief industries, providing them with a high level of physical infrastructure, but also going beyond this to support, for example, colleges to train young entrants to the industry, or exhibition centres to display the products.

Even further back behind the formal organizations, there is also the nature of social structure in the country or region. One feature of northeast Italy commented on by Fukuyama (1995) is the strong social structure in the small towns of the region. These are characterized by many formal organizations, but in addition, by many family firms, depending on extended families and able to share management functions amongst these families. At present, flexible specialization is observable in various parts of Europe, of which northeast Italy is one of the best-known examples (Goodman & Bamford 1989). In Italy there are three styles of industrial development, each typical of one region, so that the name Third Italy, for example, corresponds to a particular region and its industries.

Italy's first industrial area was the northwest, in and around the nineteenth century industrial towns of Turin and Milan, which developed their classic mass production industries from the efforts of private firms such as Fiat at Turin. A second industrial Italy was created in the Mezzogiorno from the 1950s, where the state intervened to establish iron and steel plants, petroleum refineries and petrochemical industries, again in huge plants. What has been called Third Italy arose in between, in the Emilia Romagna region (Fig. 3.4), based on many industrial firms in the small towns of the region. It is worth noting that the relevant industries in Third Italy produce items like ceramics, tanned leather and leather goods, wooden goods like furniture, textiles and clothing. Sforzi's (1989) map based on functions (specialized light industries) and social structure (industrial workers and entrepreneurs) is in fact an archipelago of industrial islands between Turin, Venice and Ancona, mostly set in a sea of rural land. Each of the towns tends to have one dominant industry, and within that industry many firms. Carpi,

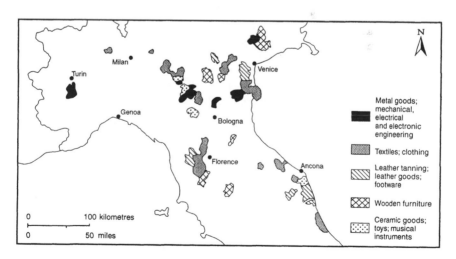

Figure 3.4 'Third Italy'. The individual small regions (mapped by Sforzi 1989) each have a dominant economic specialization. They are well defined in terms of journey to work areas, social structure (family working), and a dominant industry with many small firms.

53

for example, is a major centre for textiles and clothing, and Sassuolo is the centre for the ceramics industry.

In these industrial districts, the materials used are for the most part natural materials characterized by variability, even within one batch of timber or leather, and thus they are not readily susceptible to mass handling. In these circumstances, small industries using medium-level technology, requiring considerable personal attention to the way processes are being carried out (discard of unsuitable products, reprocessing of some faulty materials, changes in machine tensions or tolerances), have been able to survive in competition with larger ones. Many of the products are fashion goods, or at least subject to changing taste, so that the finish or the design must change from one week to the next. The firms involved are generally small and independent. Some of them are family firms, or have been in the recent past.

Among the other characteristics of the Third Italy industrial districts are: the flexible use of labour (expansion of the labour force to finish a large order, for example, by bringing in relatives of the main worker); flexible machinery, able to use slightly different materials from one batch to the next; and the flexibility conferred to firms by being in a group of similar operations, so that the loss of one market or supplier does not mean a crisis. Each firm is in touch with a number of different suppliers, and several markets, and can balance one against the others. It is worth noting that flexibility in the industrial district means between firms and products; spatial flexibility is lost, because all the advantages come from being in a cluster of firms. From the point of view of the region or town involved, this is a positive point; the industry is more likely to remain under these conditions. This happy situation does not apply to all such firms, and Amin (1989) noted another kind of flexibility in the Italian case, which is essentially a centre–periphery structure. One big firm, such as Benetton with clothing, has many subcontractors working for it, and Benetton has the flexibility of being able to change its output by changing subcontractors. The subcontractors themselves may have work only from Benetton, however, or have most of their work dominated by this one firm. This kind of flexibility is closer to that of the Japanese car-makers.

Beyond the internal structure of the firms, there is a second, social level of support for the industry, as mentioned earlier in the general comments. Communal support is available from the local chambers of commerce or industry organizations which act as information clearing houses, advising workers and firms of jobs available, and perhaps also providing technical assistance and organizing export links or other marketing for the small firms, which individually cannot manage these functions. Thus industrial districts have a logic that moves beyond straight economics. Most, if not all, the economic features could be explained in terms of external economies for small firms, but a special social structure and a cooperative attitude make the whole phenomenon more complex (Harrison 1992). It is not solely in Italy that such structures are found; to some extent, flexible specialization of this kind can be found in all of Mediterranean Europe, and examples from Spain (the shoe industry of Alicante, and the electronics industry around Madrid) are detailed in Castells et al. (1989). In some cases the flexibility is really artificial, in the sense that it is imposed by a national welfare system, and the firm's search to escape from the need to make social security payments to workers, to compensate them for dismissal or lay-offs, or to provide minimum wages. This does not render the concept inadequate, however; it is of the essence in this kind of operation that the rigidities imposed by, for example, the state or by unions, are to be avoided. In the Mediterranean cases, this has meant illegal industries. In the case of Britain, some flexibility was introduced by the British non-adoption of the Maastricht regulations of the EC.

How does flexible specialization mean an escape from the centre–periphery structure of the MNCS and their allocations of function between factories and countries? In the first place, the firms are independent, except in the Benetton type of arrangement where they are dominated by one firm which is their market. Independence means they are free to seek alternative sources for raw materials and alternative markets. Being independent may also mean more chance for innovation, although this aspect has not been tested. Independence also means a greater development of management skills and entrepreneurship, which are lost in the bureaucratic structures of the big firms.

For the labour force, such firms provide a better chance to build skills, as the jobs are usually varied, so that any worker may be required to move from one task to another as part of the flexibility of the firm. Semi-skilled and skilled labour is needed, rather than unskilled. As the skills are generic rather than specific to individual tasks, workers can also move from one firm to another – indeed, they may need to do so in order to maintain employment, or to improve their status. Because of all these features, it is possible for this kind of industrial district to arise in new regions. Rather than depend on established large cities, as in the Turin and Milan cases, or on state intervention, as in the Mezzogiorno from the 1950s, the Third Italy districts have arisen from small local specializations and skills, a distinctive endogenous type of growth which gives hope to other regions seeking industrial development.

Current changes

One of the big questions hanging over the concept of flexibility in Italy and elsewhere in Europe is whether it can survive. It may be that this kind of organization is only a transition, as social and economic evolution takes Third Italy out of traditional structures and into the modern world. What is happening now is a pull between global and local interests that may be resolved either way (Cooke & Morgan 1994). Although the region has risen to eighth in Europe on the basis of per capita income levels, with low unemployment and good growth rates for production, it may be in danger of changing to a more standard kind of industrial region.

A useful comparison of evolution may be made with a region in southern Germany, Baden Württemberg, a major car manufacturing region centred on

Stuttgart and the lower Neckar river. For this region, a principal concern is the loss of competitiveness because of the high cost of the products, partly due to the emphasis on quality of engineering, and partly to the high value of the German mark. Under these circumstances, the response by the main manufacturers, such as Porsche, Mercedes Benz and Audi, has been to use some of the principles of flexibility, but not restrict themselves to manufacturing in the home region. Porsche, as a smaller firm than the others, has always outsourced much of its inputs to the assembly line, and now uses a global search for its suppliers (Cooke & Morgan 1994). Mercedes Benz and Audi, with a heavy commitment to manufacturing their own components, have been moving towards greater flexibility by manufacturing a wider range of vehicles, and by moving out some of their component manufacture to contractors. Audi, in particular, is trying to reduce in-house production to 30 per cent of components, with the rest being supplied. Beyond this, it receives not components, but whole subassemblies or modules, so that responsibility for initial assembly is passed to their first layer of suppliers, and these in turn rely on component suppliers, who may in their turn make their components from basic parts from other suppliers.

This strategy reduces the huge responsibility and organizational requirements for the central firm, and allows more simultaneous engineering of the components, so that the problems imposed by a first design do not cripple the suppliers. But it should be noted that the industrial district strategy with home suppliers is also being bypassed, as for example by BMW, which bought the firm Rover, partly to have better access to the British market, and partly to escape the high-cost base of southern Germany. This strategy may prove more important than restructuring at home.

Large firms, heavily pressured by the competition of the small firms, and LDC competition on price, have moved themselves towards more flexible manufacturing systems. They have also taken over some of the marketing firms, so that changes in demand are to some extent under their control and can be staged and planned in advance. This applies in clothing, but also to some extent in machinery manufacture, food processing, ceramics and the rest.

Other changes are outwards movements by firms, seeking cheaper production sites in poorer regions of Italy like Puglia or Calabria, or in poor countries, especially for the basic production processes formerly done by workshops. Small firms in Italy are also under threat because they cannot easily expand, having little access to credit from banks and finance houses in a country where the capital markets are weak. Another trend is the entry of computer-aided design and manufacture. This technology is held by only a few firms which are usually unwilling to share it with others, even their own subcontractors in the business. In industry sectors such as machine tools, the problems are more severe in a way, since research and development costs are very high and cannot be met by the small firms. These must change towards cooperation with each other in order to survive.

Hi-tech flexibility: the Californian computer industry

The examples from Italy concern mostly technology which is not avant-garde but modest and well established. Flexibility can also enhance the most modern kinds of industry, as is demonstrated by the case of computer equipment in the state of California, USA. In this state, there was a modest start in electronics prior to the Second World War with the establishment of Hewlett Packard in 1937 by two graduates of Stanford University, and some other small firms linked to this university and located in the Santa Clara valley, south of San Francisco Bay. But most growth was from the 1950s with the creation by Stanford University of an industrial park, and the arrival of many established firms alongside the small local ones – established firms such as Lockheed, IBM, Raytheon and Westinghouse.

In the period from 1955, a remarkable growth of firms occurred, mostly involved in the manufacture of semiconductors, the basic building blocks of computers; 31 semiconductor firms were set up in the 1960s alone, and many of these would fail or become amalgamated with others. At the same time, a variety of infrastructural firms were set up providing services to the computer firms, such as those of finance and legal services, metalworking, and consultancies on business operation. Software firms developed which had the advantage of working closely, and often informally, with the hardware companies, and linking in easily to the academic environment of Stanford University and the Berkeley campus of the University of California. Community colleges also worked directly with some of the commercial firms, supplying them with research results and courses for employees. By 1980, over 3000 electronics firms existed in the Santa Clara valley, known since 1971 as Silicon Valley (Saxenian 1994). The industry grew in a dynamic atmosphere in which firms were created and shut down or recombined over short periods, and firms acted both in competition with neighbours, and in cooperation when advantage could be seen in the sharing of knowledge. Cross-licensing of patents to competitors was frequent, as were technology agreements to allow progress, and joint ventures which took companies forward together. Products were second-sourced (alternative suppliers found in case of difficulties with the first supplier), and this provided for more firms surviving. The corporate culture was one of horizontal linkages, and little sense of hierarchy.

Saxenian (1994) contrasts this ebullient atmosphere and long-term success story with that of the other main computer manufacturing region, Route 128, the region along the main road running north–south immediately west of Boston. In contrast to Silicon Valley, Route 128 grew up over a long historical period, from big companies manufacturing electrical goods, with 1930s links to the defence establishment in Washington, consolidated in the post-war era. Large firms were inevitable because the government required massive resources on the part of any firm accepting their contracts. After 1945, research on a grand scale was commissioned through MIT, the Massachusetts Institute of Technology, in areas such as radar, missile guidance systems, and navigational systems in general. While such research saw a downturn after the end of the Vietnam War in the 1970s, minicomputers became a success story to replace military contracts, and the industries along Route 128 employed nearly 100,000 workers in the 1970s. But the earlier source of strength in large contracts feeding large firms proved to be a weakness in dealing with the changing markets of the 1980s. Minicomputers were no longer the leading edge of technology, and in the late 1980s, over 50,000 jobs were lost, mostly in the big firms such as Wang, DEC/DG and Honeywell. The competition that defeated this region was not from Japan so much as from Silicon Valley, and the personal computer which had become the main technological solution. By 1990, Silicon Valley boasted 3231 hi-tech firms, as opposed to the 2168 firms of Route 128. In computing and office equipment, the central focus, the contrast was still greater: 294 firms in Silicon Valley compared with 120 firms around Route 128.

This is not the end of the story for the big companies such as IBM and Apple. These have learnt, at great expense, ways to find the extra flexibility and dynamism of the smaller firms. Apple, for example, has moved towards a multi-local strategy, with design centres in Europe and Asia, linked in locally to product development (prior to mass manufacture), mainline manufacturing, and the provision of services, all through local and independent firms. The comparison of the two regions does, however, demonstrate a fundamental feature of the flexible manufacturing system. It is not enough to have a clustering in one geographical region of the industry in question. There is an additional need for the right kind of firms, with the ability to network their information, staff and products through alliances, especially in industries subject to rapid change in the nature of the product.

Poor country applications

Flexibility of this kind has been studied mostly in Europe, and it is uncertain to what extent it is transferable to other parts of the world. On the other hand, flexibility is much applauded and small firms are being particularly encouraged by some of the national governments and international agencies in Third World countries. A case can be made that flexible industry is most appropriate for some of the poorer countries, on several grounds. First, the size of firms involved means that little in the way of capital investment is needed to start up. This favours capital-poor countries. Secondly, the technology involved is or can be of intermediate level, and not particularly demanding for the new industrialist. Dependency on rich countries is often because they own the latest technology, and will not give this away to poor countries. Another feature is that the industries involved, and the style of manufacturing, are heavily labour- intensive, and this also suits countries and regions with a large labour force. Because automation cannot be high in these industries, and constant human attention to the product is needed, much of the cost of production is labour cost.

An example of flexible manufacturing established in a poor country comes from Mexico (Morris & Lowder 1992). Over a third of the Mexican national shoe industry is located at Leon, a highly diversified industry with thousands of separate products and probably 5000 firms, including those of the informal sector. In addition there are over 600 firms providing leather from tanneries and components of shoes such as soles. In this industry there is a rapid turnover of small firms, mostly with under ten employees, but with an intermediate and large-firm sector providing further employment and more stability. Operation is of two main kinds. Some large firms are relatively integrated, with their own tanneries and plants for intermediate goods, and even their own retail lines. Most firms operate as flexible specialists, with small batch output that varies from week to week and requires considerable manual skills from its workers, who are usually trained in more than one department. Simple machinery allows for different ways of handling the leather. Labour is flexible in the jobs done, as well as in the hours worked; workers exhibit considerable firm loyalty so that lay-offs and extra time are possible. Further flexibility is endowed by a considerable infrastructure, with a national research and training agency for the industry in the town, and specialist chambers of commerce that seek to help in promoting and marketing the finished goods. Flexibility is also given by the way firms operate, combining readily for the handling of large batches, either subcontracting or sharing an order. The advantage of the Leon type of arrangement is that it may provide a learning environment for labour and for new entrepreneurs to build up skills and know-how in a selected industry. New industrial nodes can thus be created, separately from the global MNC structuring of manufacturing industry. It must be admitted at once that this kind of organization of production cannot be enacted for all types of industry. Highly standardized products, and those that require massive capital investment, are not easily moulded to flexible specialization. This comment would apply, for example, to the petrochemicals industry, which requires large capital investments to set up, and which produces a highly standardized set of products over a long life-time without any change.

In the best circumstances, structural flexibilities in manufacturing are combined with good access to the old centres of industry. This is the case for another LDC, Brazil, as elaborated by Storper (1991). In the outer ring of towns around São Paulo, smaller industries have sprouted up in new or expanding sectors such as electronics, and combine labour and product flexibility with good access to product information. Such a geographical location, on the edge of older industry, corresponds in a way to that described for the American automobile industry by Rubenstein (1986).

Industrial districts and regional planning

What is the significance of the industrial district for planning the development of regions? The first point to note is that the cases that have been commented on in the literature are of spontaneous growth of specific industry groups, in the sense of not being subject to a firm government policy or programme. This makes it uncertain whether such districts can be planned. On the other hand, the areas around Bangalore, in Southern India, São Paulo in Brazil, or Karachi, Pakistan, are islands of development within a matrix of underdevelopment, cases of autonomous growth with little reference to the regions around them. This seems to indicate that their growth can be established independently of neighbouring cities and regions (van Dijk 1993).

Another starting point is that there are some parallels to be drawn between the industrial district and the growth pole. Concentrated development centred on one point is vital to both concepts, in accordance with basic economic theory. For a single large industry, there are internal economies it may achieve through its scale of operation. In the case of many industries linked loosely in one sector, there are external economies. For both, there are "urbanization economies" resulting simply from being in a large city with many different kinds of industry and other economic activities.

But whereas the growth poles idea stresses the single large pole industry, firm or unit, generating growth through its own growth and innovation, and through its linkages downstream to consumers and upstream to suppliers, the industrial district concerns a multiplicity of smaller firms, and details how they act in competition and through complementarity to maintain growth.

Translating growth poles from a spontaneous process, as described by Perroux, to policy, has already been shown to be fraught with difficulties and dangers, relating to the lack of completeness in any planning of an "artificial" or planned pole, the problems of identifying an appropriate pole industry, and the long time-scale needed for the linkage structures to emerge. Similar difficulties would apply to the implementation of an industrial district. Indeed, it would be likely to be more difficult, because the key firms involved are many rather than one. There is also the point that the very nature of the industrial district is such that planning is difficult. Flexibility of labour, flexible technology, growth and decline of firms according to demand, are all problematic as planning proposals. Flexibility of labour is in part a matter of social structure; flexible technology and machinery is not the kind to be supported by standard financial backers, who cannot be told exactly what their machinery is to be used for; flexibility of the firm itself is a matter for management and entrepreneurship.

In these circumstances, the best to be offered might be supportive structures that encourage the creation of firms with flexibility. A key feature here is the environment which allows the emergence of entrepreneurs and managers, the opposite of the large firm structure in many old industrial districts, where very few workers have had any management experience. Local chambers of commerce supporting local management, training schemes and colleges devoted to the local industry, clearing houses for information on markets, and the like, can be part of such an environment.

The evidence so far

Firm structure, and still more, the arrangement of groups of firms, are seen to be of importance in the production of many complex industrial products today. Product cycles, or rather the superimposition of many product cycles on to one another as innovations, are constantly spawned from major centres of industry, and seem to indicate a continual "centre and periphery" kind of structure in industrial development. But flexible production systems suggest that this may not always be the case. In the first major example, flexible manufacturing with high technology goods and computer-aided systems leads to a kind of diffuse concentration near but not in the old centres of industry, as the US car industry demonstrates. In the second model, flexible specialization allows distinctive industrial centres to arise, even in areas of no great technological superiority, when there are many firms working in a loose cooperative structure in the same sector of industry. Such a structure may be developed in LDCs as well as advanced countries.

Beyond all these patterns, the overriding tendency, carried forward from previous phases of development, is for concentration of the assembly and producing firms in an industrial district, a 'just in time' region, or more diffuse spatial concentrations. Total diffusion out to unrelated regions, which might be in contact with one another simply by modern electronic communications systems, has not happened.

Some of the comments regarding Italy are intended as a warning about the applicability of the model of flexibility as a universal saviour. There are insufficient cases of flexible specialization to make a strong case in its favour to date, and the existing cases are under pressure to change in various ways. It may also be the case that excessive specialization itself could be a bad strategy for a region to follow. There is a case for industrial regions with a wide diversity of industries, even if some of them are poor performers at any one time. This is the view of Grabher (1993), who pointed out that those regions that have specialized heavily tend to be problem regions at a later date. This comment may extend to types of firm or of organization. If there is a place for small and flexible industries, there may also be a place for larger firms with global reach, and for different kinds of linkage system between firms. This is part of the strength of the traditional industrial regions like the southeast of England and Lombardy in Italy (Amin & Thrift 1994).

This leads us to the second section of this chapter, on the phenomenon of globalization.

Globalization

A seemingly opposite idea from that of local industrial districts and flexibility of operation is the concept of globalization: the expansion of systems of production and consumption to a world scale, exceeding that of any nation or other political unit. In fact the two are related, through the use of common devices for organizing the economy, as will be shown. But we begin by examining the phenomenon of globalization in general, as a major trend affecting even the most remote regions of each country.

For the geographer, there are several important questions about this phenomenon. What is the extent of globalization? Is it a real process, or merely an apparent one related to the pre-eminence of a few great firms, and their strength in marketing their own images and symbols? Secondly, if it is real, what are the reasons for the success of the global firm, and what advantages are reaped by it? Thirdly, what are the relations between the global firm and the locality? Is the global firm anti-local, and a threat to the locality or even the nation? Or does the global firm combine with the local and support it?

Another set of questions is posed at an even more fundamental level. Does globalization mean the end of geographical variation, the end of geography? If global firms standardize across all kinds of national and regional boundaries, what need is there to consider local markets and local preferences in production or consumption? If globalization means standardization, it may be posited that there is no need to consider regional problems at all, as they cease to exist through the permeation of all regions by the global economy.

Global and local

From the 1970s, some observers saw a change in the nature of the large corporations active in many countries. This change is encapsulated in the terms multinational corporation (MNC) and transnational corporation (TNC) (Allen 1995). Multinational companies have existed for a long time, going back to the time of the merchant adventurers organized in Spain, Italy, Britain and Holland in the seventeenth century. These companies were engaged in the control of trade between colonies and mother countries, but were also involved in the production of raw materials such as sugar-cane, cocoa and spices in the colonial territories. Modern multinationals typically have operations in various countries and for various purposes, such as extraction of raw materials, processing of foodstuffs, or the assembly of manufactured goods. They have, however, a firm home base in a mother country, where the administration and control of the company is maintained. Operations outside this country are hierarchically controlled from the centre, which is also likely to have important functions such as design and research, in the case of manufacturing companies. This is still a prominent format for the large international company, although there are variations and hybrids between it and more recent forms.

Since 1970 there has emerged a type of company, the transnational corporation, that is seen to have much less linkage to any one country. Instead, it is footloose and has no particular attachment to any single country. In order to meet the demands of different kinds of markets and infrastructure in Europe, North America and the Far East, the three main world markets, the TNC may have separate headquarters in each, and quite separate organization in each, with only a loose financial control over the whole. Any country chosen for its operations may be seen as a temporary platform, to be used while factor costs (e.g. cheap labour) are favourable, and abandoned as soon as they become unfavourable.

Up until recently, the TNC was regarded as definitely anti-local, and even hostile to national development projects, because of its lack of permanence or commitment to any country or region, its tendency to use a country for its momentary cost advantages of one or a few factors, and for its lack of interest in upgrading the local workforce skills. Holland (1976) coined the term "mesoeconomy" to refer to this new intermediate organization, an institution that lay somewhere between the ordinary firm and the state itself. Transnationals could be contrasted with the older multinationals, but still more with the more traditional, purely domestic firms of the nineteenth and early twentieth centuries, which integrated well into local economies and had greater welfare concerns for their workforce (Massey 1984). In traditional regional economies, the specialized skills of different stages of production, all put together in one region, meant a rounded set of types of work and ability. By contrast, modern concentration on one stage of production, for a foreign giant firm, means limited skill base, even de-skilling from previous levels, and a fragility of employment structure, because the single large firm might go elsewhere at short notice.

The embedded firm

Now, however, there is a new consensus that at least some of the TNCs have become linked in to local economies in various ways, not from any sense of guilt or concern for their local workforce, but as a logical means to achieving corporate aims. Some economists have seen the new emergence as an answer to the need for efficient decision-making, especially in relation to transactions, the acquisition or the sale of materials, components, information, technology, market territory, and so on. Williamson (1975, 1985) proposed that two alternative ways of handling these transactions were, at one extreme, through the market (in cases where the transaction was simple, and where the costs and benefits were well known) and at the other extreme (where there were many transactions to contemplate and each required a substantial transaction-specific investment), in a hierarchy, in other words within the firm itself. This justified the large integrated firm with its bureaucratic organization, and limited the amount of effort to be made in each transaction. Following criticism of his oversimplified version of 1975, Williamson (1985) agreed that there were forms in between market and hierarchy: various kinds of network, loosely linked organizations or groups. However, these would always merely be hybrids of the two main forms. Powell (1990) shows that in fact the network is a form *sui generis*, independent and different from the others, and having its own origins in the old trading companies, guilds and family organizations of Europe, as described, for example, by Polanyi (1957b). Such networks have the advantages described in the previous chapter, in terms of their flexibility and ability to react to changes in markets, technology, and cost of the different factors of production. They rely heavily on trust between the various elements of the network, and reciprocity of favours and responsibilities among members. While the earlier part of this chapter emphasized the small-scale firm and networks amongst craft firms, the logic applies at higher levels. For example, most of the world's car manufacturing firms indulge in network structures, as joint ventures to produce a model or a series of models in conjunction with another firm. Such firms may not abandon their traditional hierarchical structure, but enter into the joint venture for one particular market, or for one set of products.

The use of such networks does involve some of the firms in a considerable degree of embedding of their work into the locality. Sabel (1996) described such a situation for the Republic of Ireland. Here, with the encouragement of the Irish government, a variety of computer hardware and software companies, including Apple, Microsoft, Intel and Amdahl, have set up local subsidiaries, which formed strategic partnerships with many user firms and related manufacturers. The partners in some cases have become involved in the design of models for the market, pre-production development of these, and full-scale manufacture as well as subcontracting of parts to further firms. Obviously, such a degree of linkage to local firms has beneficial effects for local employment, and Sabel quotes the setting-up of firms making manuals for the use of software as an example. In such cases, a global manufacturer such as Apple uses its global reach to enhance its local connections. The whole manufacturing process, for this firm, is done in Ireland, while marketing is aided by use of the Apple advertising strength and name. This kind of development can be regarded as an example, too, of movement from economies of scale (the typical economies sought by the large firm making thousands or millions of the same object, and dividing the production system vertically from design to final assembly) towards "economies of scope". This involves production of items with a broad use across a range of models, by the small firms which are manufacturing components or services for the various computer companies, and where new designs can be implemented as a result of bringing together different combinations from the range of alternatives available. The networks allow also the optimization of choices, through "benchmarking", the identification of current best practice in the industry, and through "continuous engineering" or "simultaneous engineering", involving the simultaneous design of all the components. These replace the more traditional practices of a first general framework design, followed by design of components to fit the overall specification, a slower and potentially less efficient procedure.

Besides new working practices, networks build up different kinds of labour skills. They do not encourage the proliferation of unskilled or semi-skilled workers characteristic of the mass production system. Nor is there development of craftworking skills linked to specific materials or tools. Instead, new skills in team working, in joint problem-solving, and in group management to maximize the use of varying abilities within the group, are developed. Such skills are transferable outside the specific industry where they are developed.

In the case of Ireland, the international firms have come in partly because of the existence of public–private partnerships, the area-based partnerships, formed in 1991, which include three types of member group. First there are government agencies or departments; secondly, the firms and business organizations; and thirdly, the community sector, representing various welfare concerns. These partnerships have been important in directing the activities of the transnationals into linkage with the locals. This is not to say, however, that the industries would not have sought linkage without the partnerships aid. What is true is that these partnerships have helped Ireland become a site of successful examples of the development of the embedded firm.

There are other ways of achieving local embeddedness for the large firm. Earlier the use of the Japanese-inspired "just in time" system, with an aureole of small subcontractor firms surrounding the central one, was described, but this is a model more suitable to industrial regions within one country. A more common feature in the 1990s expansion of large multinationals is for growth to occur by the acquisition of overseas firms. Firm growth can be organic (i.e. due to expansion of the original firm itself) or through acquisitions, and the potential for rapid expansion through acquisitions has been prominent recently, partly because it allows very rapid expansion, but partly also because the acquired firms and factories are already embedded in their local economies and adjusted to local conditions. A company such as Tomkins, classified as a conglomerate in the UK Stock Exchange, has a variety of industrial products, including fluid controls, leisure and garden products, vehicle components such as transmission belts, disc brakes and flexible couplings, bakeries and food products. This wide variety means a number of very separate skill areas best handled by separate managements, and it has acquired subsidiaries in the USA, Canada and South Africa, as well as most West European countries. To expand its reach it acquired, in 1995, the Gates Corporation of Denver, a firm with 42 manufacturing plants (mostly of components such as transmission belts and hoses) in 14 countries, including Korea, Japan and China as well as Latin America: all areas unrepresented by Tomkins. Each of the subsidiaries has a somewhat different structure and relation to the local economy of the country where it operates. While exerting strong administrative control from the centre, Tomkins is able to provide a differentiated management appropriate to each locality, combining global with local.

The end of geography?

The nature of this new embeddedness gives part of the answer to the fundamental question set at the beginning of this section. There is still room for geographical variation, since the global firm is unable and perhaps unwilling to standardize its operations across the globe. One vision of standardization came from such writers as Toffler (1970), who wrote about the "demise of geography" in his book *Future shock*, but such a vision gives too much credence to the power of modern technology and high mobility to change ways of living in all regions. Technology does not have total power to change cultures and social structures that have been firmly established for centuries, and it is culture, instead, that is likely to force an adaptation of technology to local circumstances. McLuhan (1964) embraced a similar idea about the power of technology, in writing on the global village, and divided world history into three stages which were dependent on communications technology. In his first great stage, people communicated directly in speech, at a local level, in "the village". A second stage came with the arrival of mass written communications, dependent upon the invention of the printing press. This technology, allowing mass communication but susceptible to tight control over its forms and its targets, allowed a strong control and censorship of information between social groups and places. Hierarchic structuring of society and the nation state as a political device were the norm. Finally, for McLuhan we are now in the phase of the "global village", because of modern technology, using electronic means for instant worldwide communications. In the global village, as in the primitive village, society is again destructured because all information is instantly available everywhere.

Although McLuhan's predictions with regard to the technology itself are startlingly accurate, the globalization of society has not been the result, because of the prevailing enormous differences in access to communications, and the continued isolation of large groups of humanity. The model based simply on technology is inadequate because it is too simple, and from the evidence, there is a constantly shifting interplay between the local and the global.

This argument can be made more formally. McLuhan's argument is one for a determinism that runs from technology, through economy to society. By contrast, neo-classical economics assumed that the economy was independent, and that economic decisions were taken on their own. However, there has always been an undercurrent of doubt. Writers, particularly within the substantivist school of anthropology, as exemplified by the work of Karl Polanyi, regarded the economy as a whole "embedded" within a social integument (Polanyi 1957a, Peck 1994), and such views have gained recent support, as in the work of Granovetter (1985). Polanyi argued for the embedded nature of the economy in ancient societies, and thought that modern life had tended to divorce the economy from society. Granovetter makes the case that, in fact, modern economies are also embedded, and that institutions, whether firms, groups of firms in industrial districts, labour unions, or other economic groupings, all form social structures that influence the

66

way in which the economy is run. From the evidence of the operations of the modern transnationals, it is apparent that these huge firms are not immune to the patterns of social structure.

The end of the nation state?

A more sophisticated version of McLuhan's idea of the global village comes from Kenichi Ohmae (1995), in his predictions of the end of the nation state as a unit for political, social and economic organization. Drucker (1989) made the same kind of predictions, and regarded the nation state as being replaced by TNCs, by the global economy, and by new economic regions which transgress national borders. To focus on Ohmae's arguments, these seem to take technological determinism one step further, as he claims that economic structures are determining the shape of political units. He points to the breakdown of old barriers in the formation of new political units, including the Soviet Union and Yugoslavia, and the tendencies towards break-up in old established West European states such as Spain, Britain and France. Such changes are thought to be due to the effects of the global economy. This is partly reflected in the Californianization of taste amongst consumers, but it runs deeper, in the world view of people and the values they set on different goods available to them. Movement towards the uniform world view comes with the development process, and with access to expensive material goods which themselves are standardized.

Ohmae sees the changes coming about in his own culture, that of Japan, where the most global set of people are the young (the 15–25 age group). They have finally broken the web of cultural continuity with the past, and begun to question all tradition, while finding out about the outside world from such global sources as the Internet, rather than traditional sources which may be biased by local interests. Another levelling feature he discusses is the demand for the "civil minimum", for a basic level of welfare, which is found in the non-lead sectors of Japanese society such as the farming population, and in the rural regions generally. Farming in Japan has become, as in parts of western Europe, a welfare activity, subsidized heavily and totally uncompetitive on world or national markets.

These arguments can all be challenged; the breakdown of nation states is by no means a uniform process today. Artificially welded states like the Soviet Union and Yugoslavia were always likely to break up, and in states like Germany and Spain, there is strong decentralization but, as yet, no break-up of the core unit. Where there is a strong possibility of separation, as for Scotland, it is due to the long existence of a national identity not conforming to that of the state.

Nor can the changes towards Californianization of taste be agreed to be universal. While there are strong movements towards uniformity up to per capita income levels of perhaps $10,000 per annum, after that the tastes seem to diverge again, with the emergence of concern for local landscapes and settlements (each specific to a locality or region), and desire for quality-of-life elements which themselves are divergent, rather than the urge simply for more wealth. It is such trends that are documented by the move towards post-modernity observed widely by social scientists since the 1970s.

Ohmae's observations about the global views of the young can also be qualified. All young people grow old, and there is no reason to believe that the youth of Japan, like those of Europe, will not place more value on local and national traditions as they age, while their own children are seen as the new iconoclasts.

New global regions

It is possible, however, to agree with Ohmae about the emergence of a different set of economic regions. Even if we do not accept with him that "the glue holding traditional nation states together, at least in economic terms, has begun to dissolve" (1995: 79), it is evident that there are some new geographical groupings coming into being which do not conform to national boundaries. Some of the meaningful new regions are subnational, and some are transnational. It is even possible to agree with Drucker (1989) in his claim that the economy is partially organized without any direct reference to geography, by TNCs with global reach, and by the global economy itself.

The most striking of the new geo-economic regions are those crossing major international boundaries, and of course some of them owe their *raison d'être* to the boundaries. This is the case for the region along the US–Mexican border, where industries on the US side have grown up which depend for much of their labour-intensive work on partner firms or factories on the Mexican side. They benefit from lower wage costs, but are conveniently centred within the USA and have good access to its markets. A comparable case is that of Hong Kong and the adjacent parts of southern China, in Guang Zhong province, where China itself has set up major industrial towns producing for Hong Kong companies using very low cost Chinese labour, instead of bringing that labour to Hong Kong itself, as happened through the 1950s and 1960s. The region of Hong Kong, Shenzen, Guangzhou, Amoy and Zhuhai is already a major global manufacturing region. Another such region, with its focus on Singapore, includes the Malaysian state of Johore and the Riau Islands of Indonesia. Global regions within individual states are cited too, one being the Mississippi Valley, with its rapidly expanding set of car manufacturing plants, many of them having Japanese owners, but feeding into global markets. Another global region, itself having a GNP that is second only to the whole nations of the USA and Germany, is the Shikoken region between Tokyo and Nagoya in Japan.

It should be noted that the new global economic regions are not necessarily tied to the operation of the footloose TNCs. The reason such regions exist is to do with spatial permanence and an investment in a particular geographic space. Many of today's international companies are in fact quite traditional MNCs, with a

68

continuing power base in one country and subsidiaries outside. Thus, for example, Matsushita, Japan's, and indeed the world's, largest electronics company, remains firmly based in Japan, despite recent moves to open more overseas factories and functions. A quarter of the group's production is now overseas, with plans to expand to 30 per cent by the year 2000. It has large investments in China: up until recently only assembly operations for items such as air conditioners and television tubes, but now including research laboratories. Matsushita now has ten such laboratories overseas. However, central control and most manufacture remains in Japan. Expansion of all activities is predominantly into China and other Far Eastern countries, which form a growth region with the Japanese core.

While all these new industrial regions can be identified as real economic powerhouses with real linkage amongst their parts, it still makes sense to discuss countries as units for development. Apart from the cultural integument to each national economy, there are direct economic differences at this national geographical scale. Labour skills and labour costs, the levels and availability of technology and resources, are factors that vary more on the between-nation scale than in any other dimension. But as is shown elsewhere in this book, the cultural integument is important too. Economic growth depends on a stable governmental structure, and in the long term, one that is democratic and does not invite revolutionary change. It is also dependent on the absence of coercion within the state or within institutions, and the absence of corruption, which is linked to overpowerful and illegitimate systems of governance.

There are also changes within the advanced nation states which give them continuing advantages. As Ohmae (1995) comments, there are specific types of response from the nation state to the crisis of change. Changing demands within these countries create opportunities for new employment, especially in the service sector of the economy, in industries such as tourism, recreation, and the burgeoning business services, and these kinds of activities take over the central position held formerly by manufacturing industry. The "hollowing out" of the industrial economy in the developed countries does not mean that these countries are left with no function, although old industrial regions may experience major trauma in achieving the changeover.

CHAPTER 4
Social components of development

As development in the last 50 years has largely been seen as economic develop-ment, it is unremarkable that social aspects have received relatively little attention. Social change has been analyzed by sociologists in a discourse separate from that of development. Social elements in development have been treated as a policy matter, covered generally by welfare. This lays immediate emphasis on con-sumption of social goods and services. In other words, society has been regarded as the consumer, the recipient of development, and not as the producer. In the following discussion, we examine first the place of social components in development, and then review developed countries and less developed countries for relevant materials.

There is no common agreement on the place of social elements in the development process, so that it is worth outlining briefly some points of view; whether this social component should be provided for in policy by public agencies or be left to the private sector is another debate. A first viewpoint on social provision which we might note is that of the psychologist, Maslow (1954), that human needs are ordered in a commonsense way, so that survival or economic interests, food, shelter and clothing, are first, and only when these are satisfied do social provisions come into focus. Social facilities are secondary and thus come in at a later stage of development. The logic of this view is that social concerns are worth including in development policy but not in the early stages, which means amongst the poorest countries.

A contrasting view is that of the economic historian Polanyi (1977), who showed that historically, social concerns are in fact primary, even in the most primitive peoples, and that the economy works within a social framework, what he called the "social integument". In tribal society, economic work is allocated to specific persons, at specific times and places, set by the community. In recent decades there has been an attempt to treat the economic as though it were autonomous, but this is a mistaken procedure. If we take Polanyi's view, the structure of society is of fundamental importance at any stage, and should be addressed by policies seeking to promote the overall development of any human group.

A third view, implicit in neo-liberal economics, is that the economy is entirely autonomous and its development may be pursued as a separate enterprise, whether or not social concerns are important to the country involved. This kind of thinking might be thought to underlie some of the policies of present-day Western states, and certainly ties in with the earlier economic development writing (see the selection of items in Agarwala & Singh 1958).

On the other hand, a leading writer on liberalism in the modern age, Friedrich Hayek, although he rejected the welfare state, did see a need for some level of social provision. To Hayek (1986), the creation of the welfare state in Britain after 1945 constituted a "road to serfdom", paved with good intentions but having disastrous results. Anything that distorted market forces created an artificial environment, making both people and whole nations maladjusted and thus unable to compete and to earn the income with which to provide their people with benefits. However, he saw a need to provide a base level of welfare – some people in some circumstances would need the humane provision of some services to avoid destitution.

In recent years what might be regarded as a fourth view, or a re-emergence of older thinking, is that the social structure is of considerable importance, and a strong differentiating factor between different countries and regions. This thinking comes from diverse sources. On the one hand, the post-modernists or post-structuralists (Shuurman 1993, Peet & Watts 1993) see a need to engage with the diversity of experience in different societies, and to move away from a standard account of how development happens. Each society has a different development agenda, and any central power that wishes to impose its own programme must adjust to that society. On the other hand, Fukuyama (1995) calls attention to the strength of local social organizations in influencing the kinds of firms that come into existence and which flourish in particular societies. In between these, another line of thinking emanating from the Social Market Foundation is that, apart from pre-existing social structures, it is always of value to improve the human capital of any society, and that most of what we term development depends on this high-quality "human capital" once the first stage of development based on simple unskilled labour is passed.

In this chapter, the position that social provision is an integral part of development and cannot be set aside will be accepted, on the basis of empirical evidence rather than theory. Some evidence for that position comes from data showing that social provision is generally higher among the more economically developed countries. However, this does not prove any relationship and the causality might be the opposite way, in that economic development allows and promotes better social provision. A better case for the active role of social development is made through seeing the "human capital" of skilled and healthy workers as a factor in development.

Developed countries

In this section, rather than describe the levels of social attainment generally in developed countries, we focus on a broad question, that of the relation of policy to social development. Has the strong provision in Western countries been useful to overall development, or might it be that some elements of social structure have actually been damaged by the welfare state?

A central question within the more developed countries is not the reality of social change or social development, which may be measured in terms of education levels, housing or medical services, but whether this social provision is best provided, directly or indirectly, by the state, and whether the state is able to direct its efforts equitably or efficiently. Evidence shows that in those countries where welfare has been provided, it has tended to expand to cover more than basic provision, and to become ever more complex. In the UK, over 90 per cent of education is provided by the state, over 90 per cent of medical care, and before the housing reform of 1980, over 30 per cent of housing. There has been a concern to provide the welfare as evenly as possible, indeed to use welfare provision as a means to reduce differences between the poor and the rich. In the event, this has not been possible.

We may take three examples. In Britain, educational facilities, especially those for higher education, have traditionally been taken up more strongly by the middle classes, those most aware of the value of education. The provision of secondary schooling, and still more of universities as a free but voluntary service, has been of greatest value to these groups, while paid for out of general tax revenues of the country (Marsland 1996). As another example, the provision of public housing in Britain has been through local government, which has run this service in an uneven fashion. In Glasgow and most industrial cities over the 1930–80 period, the council house provision was via a standing list, the top people on the list being those who had been on it longest. This favours long-term residents against newcomers, those who are immigrants to the country or to the city, or newly-formed families who may well be the people in greatest need of housing. A third example is in the take-up of healthcare, which Le Grand (1982) has shown is highest for the middle classes in Britain. In one study, professional people had a 40% higher expenditure on healthcare than unskilled manual workers, partly due to stronger perceptions of the need for care, and partly to greater ease of access (private cars, time available) to the services.

These three examples indicate the variety of irregularities in provision and the complexity of any attempt to address the issue of equity. But there are other criticisms of the welfare provision which may be more damaging. In the examples given above, the role of the provision is liable to work against economic forces and to reduce the efficiency of production in direct ways.

Thus, housing provision in Glasgow has created a barrier to the flow of labour from the city region to other areas of Scotland or the UK over the last 40 years, which have seen constant job losses in the traditional industries of steel, shipbuilding and heavy engineering. In reality there is a need for some labour movement into other regions and other industrial sectors, to loosen labour-market rigidities. For education in Britain, the heavy take-up of higher education by particular socioeconomic classes is likely to exclude some of the most able potential scholars, and thus lower the overall level of academic attainment.

At a general level, there are other criticisms, including the creation of an "underclass" of welfare-dependent people, especially in those regions where welfare payments are largest, and the inefficiency of a monopoly provision of welfare with no competition (Marsland 1996). These negative features are strongest, or at least most readily seen, in countries such as Britain with a long welfare history. Perhaps the strongest negative criticism concerns the destruction of social structures which were in place before state intervention in social matters. This is the topic covered by Francis Fukuyama.

Trust and the social virtues

Rather than the creation of a dependent underclass, Fukuyama (1995) has focused on the destruction, in some modern societies, of the existing social structures that stand between the individual and the state, and which are necessary for the proper functioning of economic life. He shows that while the standard view of societies is either as individualistic (such as the USA) or as communitarian (such as China or Japan), there are in fact many societies with strong social structures at the intermediate level. These are seen as the most successful societies, ones in which economic transactions are conducted with ease because of the trust existing between the opposing sides in the transaction, and the sharing of information within and between pre-existing social structures.

It is worth listing some of the different kinds of case involved. Japan and the USA, in fact, are portrayed by Fukuyama as strong civil society countries, where there are many organizations at the intermediate level (families, firms, societies and clubs) not dependent on the state. The cases where there has been a very strong state organization of society in the past, such as the USSR and southern Italy, and where this has now broken down, have the greatest problems because their economy is liable to fall into criminal hands, for want of an alternative set of institutions. The Mafia of southern Italy, a criminal organization filling the vacuum of legal organizations, is paralleled today by Russian "mafias". Cases with a strong family social structure include that of northern and central Italy, home of the extended family. This structure is valuable at early stages of economic development, but imposes difficulties at later stages when the firm or organization becomes too large to be handled by one family. Trust in such cases has to be extended, and may cause disasters if the trust is not justified, or if incompetent family members are chosen instead for important posts.

For countries that have suffered destruction of their earlier civil society, like the Republic of China, France, and southern Italy, there are major problems in economic development because the state has to step in to initiate programmes, and often makes mistakes for want of sensitivity to markets and the interests of society.

Throughout Western society, however, there is a tendency for civil society to break down, because of the very strength of social provision by the state. For Fukuyama, again, this is the result of 200 years of social engineering, since the time of the French Revolution, and the welfare system is counterproductive to the wellbeing of the people. This occurs, not just at the social level, but also through its effects on firms and transactions at the level of the economy. It is, of course, false to ascribe the breakdown of social structures entirely to state intervention. Migration from village communities to cities, and the technology of modern living, must have important effects in the same direction. But it is notable that some countries appear to have been able to maintain strong civil society while undergoing urbanization and industrialization.

This whole field has been left aside by geographers. There are no geographies of the degree or intensity of social interaction or strength of civil society. It is likely, however, that this will be an important field for the future, because there are obvious and important differences between regions and between nations in the kind and strength of the institutions at the intermediate level.

Society as a productive force

From the previous discussion it is evident that there are great irregularities in the way welfare provision is distributed and actually taken up. Although most people would agree on the desirability of social provision, it has proved exceedingly difficult, even in advanced industrial countries, to make public provision in an equitable fashion. In addition, the welfare system in Europe would seem to have created a dependent group or underclass, while social cohesion and civil society has deteriorated, in part due to the rise of the welfare state. If we move from actual processes of social change to policies, it is uncertain what can be done to support civil society, and one inference may be that the state needs simply to reduce its intervention. There is one broad area where intervention can be significant, however; this is in the field of improving human capital, of producing a society able to work in new fields and with useful skills.

Rather than social justice, the emphasis moves to improving the human resource for production. Perhaps the most central elements in this provision are education, training and retraining. If we consider the human element in production in terms of people with skills, not as a homogeneous labour force as did both the classical economists and Marx, then it is possible to conceive of human resources that are created, improved or destroyed over time.

This takes the viewpoint entirely away from the moral arguments for welfare, commonly adopted to support the welfare state within a Christian culture, in which the idea of equality is dominant. From what may be called a supply-side view, good provision of education at all levels is necessary simply to raise productivity, alongside provision of other services such as housing or medicine in cases where the market fails to provide, in order to sustain an active and satisfied workforce.

Regarding labour training and skills as a "resource", or as "human capital", obviously makes it a more acceptable item for investment in schemes of economic development. For example, this view is taken by the European Union's Social Development Fund, one of the structural funds directed at specific areas of the Community. Putting social development of this sort into policies means two kinds of action, both of which have important geographic elements:

(a) identifying and aiding areas with declining economies, where substantial skills have been built up, where a culture of organized and skilled work is in place, and where retraining is possible;

(b) building new skills amongst people previously without training in modern types of employment.

Taking the first of these, retraining is seen as a worthwhile investment in many old industrial areas. From the point of view of the region and its inhabitants, there is interest in training schemes, especially where there was previously an excessively narrow economic base and the skills developed over generations are not really transferable. Such is the case, for example, in European coalfields and steelmaking regions, where the skills of the workers are so specialized as to be almost useless in other skilled areas. Other cases include most of the industries from the two earlier Kondratiev industrial waves: textile manufacture, heavy engineering and shipbuilding, and basic chemical industries. In Britain, most of the coalfield regions fall into this category, notably those with little diversification such as the South Wales coalfield with its coal and steel, or the Yorkshire coalfield with coal and textiles. There is no adequate statistical index that shows this structure of industry, although the geography of the coalfields itself, as this shows the basic resource, is a fair indicator.

On the other hand, from the point of view of incoming firms, or even of the nation, these are not attractive regions. The industries in them, in most European cases, have been owned by the state or by large monopolistic companies for several generations. Because of this they have been operated as non-competitive monopolies and in an inefficient manner, and their workforces have often developed monolithic union structures, unwilling to accept change. They are manifestly not success areas, and represent past industrial revolutions, not just in technological terms but also in their industrial culture. They can also represent a labour force geared to past situations, as Drucker (1985) shows. While their education and aspirations remained little changed over the twentieth century, what has changed is their relative wage rates, so that they have become used to the high wages commanded by the power of their trade unions.

Regional development agencies in these regions need to work at changing the public image of the area they represent, through city or region marketing, and through defining areas of high quality of life. As for the workforce, they need to concentrate on retraining the workers who have lost employment in the old industries, and provide new education for the next generation of workers.

An example of the attractions to firms is presented by Scotland. The Central Scotland belt is currently successful in attracting new employment and some new firms, especially in the electronics sector. The new firms are locating, not exactly in the main sites of the old industries such as Glasgow with its multiple heavy industries, Motherwell with its steelmaking, Clydebank with shipbuilding, or Dundee with its jute mills, but in small and medium-sized towns on the edge of the region with more attractive environments for a new white-collar workforce, replacing the blue-collar force of the coal and heavy engineering industries. Only where very special credits and aid are given, as at Motherwell and Clydebank with Enterprise Zone status, do firms reoccupy the old sites.

In the new Scottish regional development agencies, there has been some success in combining good quality-of-life elements as attractions to firms, with a positive programme for developing human capital. Thus Scottish Enterprise, the main agency for the nation, distributed in 1994, through its 14 local enterprise companies, 37.5 per cent of its funds into education and training for youth and adults.

To take a single one of these companies, the Skye and Lochalsh Enterprise, based at Portree in Skye, in 1994–95 put its efforts into many human capital investments. One sector was its programme for Enterprise Training, helping individuals to set up firms. Another two were Training for Work, and Youth Training, helping employees, especially new trainee entrants to firms, to get the necessary skills. One was for Community Action, helping local communities to organize themselves and strengthen the community structures. Attached to this is a programme for revitalizing cultural heritage, especially the use of the Gaelic language, in recognition of the fact that traditionally strong communities were found among the Gaelic-speaking communities of Skye. Apart from individual schemes, the style of the agency is to work in partnership schemes wherever possible, relying on and building local communities or firms as social units which will make their own decisions rather than grow in dependency.

The Skye agency's success is hard to measure in precise terms. An indirect measure is the growth of population, of over 10 per cent in both the 1970s and 1980s, following a long decline in earlier decades. This was not a growth of retirees, but of young workers for the most part; the employment growth over 1981–89 was the highest in Scotland at 61.9 per cent (Scottish Office 1992).

Less developed countries

For these countries, rather than examine directly the welfare issue of the developed countries, we look at the empirical evidence of social levels in various countries. These tend to show a strong positive relationship between the social and the economic, although the direction of causality is not clear.

Concerning the civil society and Fukuyama's claims for its importance in development, many LDCs have an advantage in that their traditional civil society base, the family and the extended family, as well as village communities, are often still in existence and have maintained strength despite the onslaught of modernization. On this strong social base, such countries have been able to make good use of the early stages of welfare, the provision of primary education facilities, of basic medical aid, and of housing for the needy. In these countries, welfare is not destructive of society but supportive, so that our conclusions as to the role of social programmes may be quite different from the preceding discussion.

Another point to be drawn from the data presented on Brazil is that there are strong regional patterns within countries, and many remote or inaccessible regions have difficulty in receiving the welfare on offer.

Empirical evidence

Because of the uncertainty as to how social provision works, and the direction of causality, it is worth detailing some of the levels of national provision in individual countries. A complete survey cannot be attempted, but two sectors, education and health provision, are examined below.

Education

From Table 4.1 below, it becomes evident that there are major differences between the LDC groups, and that South American countries compare well with sub-Saharan Africa.

The major achievements since 1965 in all countries are apparent. But there are substantial gaps between each level in the income table. Low income countries, particularly when China and India are excluded, have improved from a very low base in 1965, but still have only 26 per cent of the relevant age group in secondary

Table 4.1 Educational levels in LDCs compared with other countries, by percentage of each 12–18 year cohort in secondary education.

Group	1965	1987	Female 1987
Low income states	20	37	29
China & India	25	41	33
Other low income	9	26	16
Low middle states	23	49	50
Upper middle	32	67	65
High income	62	93	96

Source: World Bank Development Report (1990). Low income, under US$545 per capita income in 1988; lower middle, £545–2,200; upper middle, $2,200–6,000; high income, over $6,000.

education. In these low income countries, there is also a large gap between male and female education levels, with only 16 per cent of girls receiving secondary education.

As the newly industrialized countries (NICS) are often mentioned as evidence of the value of education, some representative values on the same parameter, percentage of the relevant age group in secondary education, are given in Table 4.2, including some NICS-in-waiting. These illustrate high levels of education by 1987, but more dramatically, the rapid rise of education provision in South Korea and Hong Kong over the period. These countries have gone through more than an economic development push; they have achieved social development at the same time, and the data illustrate a synergy between the economic and social aspects.

Figures for two sub-continental areas illustrate further the range of education provision. Data for South America and for a range of countries in sub-Saharan Africa are shown in Table 4.3. South America stands out remarkably well against Africa on this evidence. The lists are not complete, but the range is representative. In this case, the comparisons are coloured by the fact that the South American countries have a much higher level of urbanization. There are more schools and

Table 4.2 Education levels in NICS, as a percentage of each cohort in secondary education.

Country	1965	1987	Female 1987
South Korea	35	88	86
Hong Kong	29	74	76
Indonesia	12	46	–
Philippines	41	68	69

Source: as Table 4.1

Table 4.3 Percentage of relevant age group in secondary education in two continents.

South America	%	Southern Africa	%
Bolivia	37	Mozambique	5
Ecuador	56	Ethiopia	15
Colombia	56	Tanzania	4
Paraguay	30	Somalia	9
Peru	65	Burkina Faso	6
Chile	70	Uganda	13
Brazil	39	Sudan	20
Uruguay	73	Togo	24
Argentina	74	Côte d'Ivoire	19

Source: as Table 4.1.

places for pupils in towns, and because of the organization of urban life, schooling is possible since fewer children are taken out of school to help on the farm.

Education for South America seems to be both a result as well as a cause of the central industrialization–urbanization process in development. This warns us against any easy assumption of causality between education and the general process. Latin American countries have been able to advance educational levels, but have often been unable to provide employment for the best educated people. Through the 1960s and 1970s, Argentina was notorious for exporting large numbers of professionals to other countries. This "brain drain" was only the most extreme element of an imbalance between educational achievement and economic possibilities.

Health provision

The patterns for health provision are similar to those for education. One index here is the incidence of infant mortality, relating to overall health provision, health conditions for the mother and for the child (Table 4.4). It is chosen here in preference to food provision indices such as calorie supply, which do not measure the quality of food provision; and to the numbers of doctors, which may reflect conditions such as those for Latin American countries where too many professionals are produced by the educational system.

Again comparing South American countries with those of sub-Saharan Africa, the American range is 23–108, versus an African range of 41–168. As with education, the urbanization factor comes into play, as well as local emphasis on

Table 4.4   Levels of infant mortality.

Country	Infant mortality per thousand births
India	93
China	31
Other low income	98
Lower middle	57
Upper middle	42
High income	9
NICS	
Hong Kong	7
Korea	24
Emergent NICS	
Indonesia	68
Philippines	44

Source: as Table 4.1.

particular health provision systems, but the broad difference between continents is apparent.

Social and economic balance

Given that in the NIC cases mentioned above, the provision of social welfare has been stimulated or managed throughout by the state and has accompanied a strong development process, it may be asserted that social provision is of positive value. It must be emphasized again, however, that a balance needs to be struck. There are cases at the national level of social provision which runs ahead of the economic prosperity of the country, so that imbalances occur. In particular, health and education provision in countries that cannot offer enough employment can have a negative overall effect; such imbalances are to be found in countries such as Cuba and China, constrained by political systems, and in some ex-colonial territories such as Sri Lanka.

In Sri Lanka there is 100 per cent enrolment of pupils in primary education, as against typically 20–50 per cent in "low income" countries on the World Bank scale. Life expectancy is 72 years, against an average of 60 years in low income countries, infant mortality 21 versus 72 per thousand, and adult illiteracy 13 per cent against 44 per cent. High levels of education and social provision generally, in the absence of jobs, can be an explosive political mixture.

Regional patterns: Brazil and Ecuador

Within a country, there can be similar problems at a regional level. Taking the example of Brazil, the Nordeste region is the poorest in the country. Per capita regional product figures, shown in Table 4.5 (in thousand cruzeiros), are six times higher in São Paulo state than in Piaui in the Nordeste. Welfare provision, as indicated roughly by the listing of sewage disposal facilities (Table 4.5), is also far lower in the northeast than in the southeast.

As an education indicator, we may use the number of pupils finishing secondary school, in relation to total state population (Table 4.6). For the major regions there is not too extreme a difference, although the southeast has far higher numbers of "finishers" than other regions, despite an older age structure. But there are major local differences in the northeast. On top of these differences, the school finishers are almost all in the urban areas of each state. Secondary school graduates in the rural parts comprise less than 2 per cent of the small percentage of children actually receiving secondary education. Education provision of any kind is scarcer in the rural areas, which have always been the periphery of a periphery.

But the provision of improved facilities in the Nordeste must be set beside the present-day tendencies. There is heavy underemployment, especially in the rural areas of the region. Mass migration out of the region, towards the southeast,

Table 4.5 Measures of wealth and welfare provision in regions of Brazil.

State or region	Regional product per capita (1000s cruzeiros)	Sewage disposal: % of localities with service
Nordeste region	4,373	26.1
Ceara State	2,514	7.6
Southeast region	12,386	91.0
São Paulo State	14,515	94.8

Source: Anuario Estatistico do Brasil 1995, Instituto Brasiliero de Geografia e Estatistica, Rio de Janeiro.

Table 4.6 "Finishers" of secondary school as a percentage of the total population, in various regions of Brazil.

State or region	% finishers
Nordeste region	0.32
Ceara State	0.003
Rio Grande State	0.52
São Paulo	0.54

Source: as Table 4.5.

involves mostly the young and the better trained segments of the population. Any one-sided improvement in education, health, or other social facilities, is likely to increase the flows to the southeast beyond the level at which they can be absorbed into the cities. It is unlikely to attract in new industries seeking to utilize the skills of the population. As in the national case of Sri Lanka, social provision cannot be justified as the lead sector of development.

The Ecuadorean case provides a sample at a more local scale. Teachers and doctors for the rural region around Cuenca, in the southern highlands of the country, mostly live in the city of Cuenca and travel out from the city to visit their work areas. At the village of Sig Sig, some 40 km from the city, for example, there are some primary school teachers, but the secondary school is taught largely by teachers who take buses out from Cuenca every day, or in a few cases, by teachers who arrive on Monday and leave on Friday. Doctors are only available here because in the first year after graduation, a field year is imposed on new doctors who must work in villages. Again, they are available only during the week. Smaller villages around Sig Sig have no service.

Human capital

The above review of basic facilities has avoided the ideas presented in the first half of the chapter, on improving human capital by upgrading skills and attracting high-technology industries, as has become the strategy in some developed countries. This would seem to be less relevant to poorer countries where the industrialization process starts with advances demanding little in terms of skills.

However, if we look at countries such as Taiwan, it becomes apparent that human capital has been improved and upgraded continuously in recent decades, because the country has moved from a farm goods exporter, to a manufacturer of textiles and plastic goods, to electrical and electronic goods, in each case demanding higher skills. What has happened is a process of in-house training within firms. This is a model for upgrading that has typified development in the East Asian model, starting with Japan and continued by all its followers.

In general, the case may be made that initial provision may be from the state, but that the baton should be taken up early by the private sector, matching the economic policies.

Conclusions: natural process or welfare policy?

An overall conclusion of the chapter is that social components certainly exist within a development process, and that there is a two-way positive feedback between them and economic aspects of development. In the developed countries, however, there is the potential for a number of negative features to appear as a result of social welfare incorporated and carried through over many years as a formal policy. This includes the decline or destruction of civil society structures, such as the family and voluntary associations of people outside the state organizations, a result that is the very opposite of that sought by policy. In this way, economic development may be accompanied by social decline, or undertaken at social cost, in the same way that environmental costs exist for all economic development.

In LDCs, however, the case is different, and state intervention to promote welfare is probably justified, since the traditional structures of civil society are not broken down by the basic forms of welfare which provide only for the needy. The example of countries such as Taiwan is that, as with economic development, social provision should move from state support to the privately sponsored at an early stage.

It is clearly the case that the state, in some of the poorest countries, is actually unable to make any provision, and in such cases a variety of agencies, often foreign, may come in to help. These are the non-governmental organizations or NGOs. A tremendous weight of emotional feeling attaches to the provision of welfare to the poorest countries and regions. The work of OXFAM or CIIR (Catholic Institute for International Relations) and other NGOs has long been to alleviate all long-term suffering through poverty in these regions. A large amount of this work is in fact in rural areas subject to major physical problems – the climatic hazard of drought in the Sahel countries, flood liability in Bangladesh, and drought plus a dense rural population in the Nordeste of Brazil.

These efforts are not above criticism. In such areas, the NGOS are beginning to realize only now that helping the people simply to survive in such areas, by providing famine relief, is creating or maintaining the problem rather than solving it. Providing medical services often starts with improving sanitation, and the effect is to reduce infant mortality and help the survival of infants under one year old. This increases population pressure on a limited resource of land and water. Birth control, a more significant long-term approach, has proved difficult to enact in Asia and Africa. In Latin America, it has been positively resisted by people since it is not endorsed by the Catholic Church. Coming as an initiative of outsiders, it is also seen as an imposition of the rich countries on the poor – the rich wish the poor to limit their families so that more surplus is available for export to the rich.

The ideas of balance presented earlier mean that improvements in medical services of all kinds are best introduced alongside economic improvements; economic improvement in line with social service improvement means that extra production can support a larger population and the demand for more services. In the event this is what tends to happen through the transitions described earlier, transitions that occur universally and spontaneously, without the need for special institutions to carry them through. Demographic transition to low birth and death rates comes with increased urbanization and industrialisation of the country or region.

CHAPTER 5
The service economy

An important part of modern economic development is within the service sector, including retailing and the sale of goods to the public, but also the provision of services of all kinds, some of which were treated in the previous chapter as social provision – education, health, and welfare in general. Another set of services for individuals and families is in retailing, and the provision of leisure services, recreation, health and beauty, and tourism. All these are experiencing growth in countries with advanced economies, but a still more central feature to current development has been identified as the services provided for firms, called business services or producer services.

Services as an economic sector have had little treatment by students of development. This may be attributed to a central tenet of socialist thinking, from the late nineteenth century, that the production of goods was the only kind of production with real value, and that services were purely accessory. Viewed from this angle, it is only necessary to plan manufacturing production and the rest will follow. Another point is that most development writing has been about Third World countries, and it is certainly true that for poor countries, service-based development has not hitherto provided a promising base. With the rise of the electronic movement of information, this is already changing and will continue to do so in the future.

Services to firms have a particular importance because they link to manufacturing generally, and because some kinds of firms make particularly strong use of services. Global organizations, operating in several continents, choose as their regional bases, to act as centres for operations, those cities with a wide range of professional and other specialist services on which they can call. Often the specialist services have grown to match the growth of the global firms.

In a different kind of case, some industries are composed predominantly of very small firms, and rely heavily on the provision of services as external economies: services that they would otherwise have to provide themselves at higher cost, but whose cost is best shared with other small companies. These concentrations of business services have their own geography, linked to that of the relevant industries.

This sector of the economy is taken separately in this chapter, not just because it involves different activities with a different geography, but because the most rapid economic growth in advanced countries is in this sector. As statistical data show, employment shifts continuously into services in the advanced economies.

We will concentrate, therefore, on these advanced economies and their service sectors; there are large service sectors in many poor countries, but these do not generally include service industries producing for export markets and driving the respective national economies. Often, in fact, they are of low productivity in the informal sector in the cities, serving as provider of poorly paid jobs for those unable to gain full-time regular employment. This does not negate their important social role, often as a training ground for migrants from the countryside who are able to learn urban ways through working in these activities, but it does exclude them from a central developmental role.

As Table 5.1 shows, there is a normal course of events followed by most countries over a period of time, which sees the transition from dominance of the primary sector, to the secondary sector, and in the later stages, to the tertiary and quaternary sectors. Some countries certainly do show a tendency to delay their manufacturing decline, such as Japan and Germany. These countries have been particularly well endowed with human and physical resources for a manufacturing role, and have certainly been able to manufacture successfully for world markets from a home base. It is none the less the case that even in such countries, the role of manufacturing, in employment and also as a proportion of GNP, declines eventually and is balanced by a concomitant rise of the service sector. This deindustrialization shift can be seen at the regional level too. In the USA (Rodwin 1989) the decline of industry was first observed in New England and the northern Appalachians, in the 1950s and 1960s. Later, the 1970s saw decline of manufacturing in the middle Atlantic states, Virginia, Maryland and Delaware. In the 1980s, the change was affecting the industrial Mid West, in such states as Ohio and Michigan. This "hollowing out" of the North American industrial regions has generally been viewed with alarm, although Rodwin shows it to be a predictable process.

Table 5.1 Employment in manufacturing as a percentage of all employment in various developed countries by year.

	1955	1966	1981	1992
Austria	29.8	29.8	29.7	25.6
Finland	21.3	22.8	26.1	19.3
France	26.9	28.7	25.1	25.5
Germany	33.8	35.2	36.7	33.6
Italy	20.0	25.8	26.1	22.1
UK	36.1	34.8	26.4	26.2
New Zealand	23.7	25.4	25.7	16.4
Australia	29.6	28.6	19.4	14.5
USA	28.5	27.8	21.7	17.0
Canada	24.1	23.9	19.4	14.6
Japan	18.4	24.4	24.8	24.4

Source: OECD reports, UN Statistical Yearbooks.

There are different explanations of this deindustrialization. A "post-industrial" view is that manufacturing declines: (a) because of higher productivity and capital investment in this sector compared with services (in other words, manufacturing does not create jobs although industrial production grows); and (b) because of a constantly expanding demand for services, compared with manufacturing, at higher levels of income – the economist terms this a higher elasticity of demand (Rodwin 1991). Another interpretation is in terms of the expanding role of the multinational corporations, which can shift investments rapidly out to the developing (and cheap) countries, leaving central countries with just the firms' head offices. Yet another explanation is in terms simply of the shifting balance of comparative advantage for trade, so that the LDCs become better able to compete in manufacturing and cause it to shift out towards them (Beenstock 1983). This is really the same concept as the MNC argument, but not focusing on the firm. In the USA, the shift has also been interpreted as a shift from the "Rustbelt" (the name given to the cold northeast because of the predominance there of older industries, many of them based on iron and steel and now literally and economically rusting away) to the "Sunbelt": a move from harsh climates and industrially polluted landscapes, to areas with better quality of life, such as California and the South (Glickman & Glasmeier 1989). A similar explanation can be given for the rise of industry along the Mediterranean coasts, such as that of Spain (Morris 1992b).

In any case, there seems to be no reason that many of these shifts should not be permanent. What the movement out of manufacturing industry does leave behind is a well of human skills, of organizations, and in fortunate cases, of institutional and company bases and home offices, which do not disappear when lower cost labour for simple operations can be found elsewhere, or when a physical resource is exhausted.

Services as an engine of growth

In the modern world, of course, the production of services, personal, business, financial and administrative, is considered of equal value to the production of goods. Under the old thinking, a manufacturer of rubber tyres was making something that could be sold abroad and earn profits for the company and foreign exchange for the country. The banker who supplied the tyre manufacturing company with credit for his export operations was an accessory to the central function. Today, in many West European countries, the service sector, especially such activities as banking, are working as the exporters, supplying credit and financial services worldwide, while such goods as rubber tyres are accessories which are used by the workers and managers of the financial sector. To use the terminology of industrial economics, the competitive advantage of some countries (Switzerland, the UK), and of some regions or cities (Zurich, London), is in service provision.

Under the old thinking there was a special geographical pattern to industrial location, while services might be spread quite evenly across the country, simply matching population distribution. Thus, for example, there was a precise geography to iron and steelmaking because of the high transport costs of the materials from which the iron was made. Steel manufacturing was limited to sites near or on coalfields and iron ore and limestone supplies, or later, on coasts where these materials could be assembled. Banking services, by contrast, might be found in every market town with a few thousand inhabitants. Central place theory did serve to highlight the hierarchy of services, with lower services in small places, and higher order ones only in the largest cities. But this was a size ranking of services, placing high-order services in the centre of large market areas or large populations, and low-order services in the centre of small areas. There was, in most cases, little need for these services to agglomerate for interaction amongst themselves, nor for them to locate in reference to raw materials, power supplies, special skills in a labour force, or any other special factor. Now, however, there are major concentrations of services, some of them new ones catering to new industrial structures and needs, such as advertising and marketing.

Global cities

In Chapter 2 the world systems concept of the global city was briefly described. This concept is relevant to the geography of the producer or business services. King (1990) and Sassen (1991) have analyzed the structure of some of these cities, which have emerged to cater to the new needs of global companies.

To take an example, the financial role of the global cities is indicated by the presence of foreign banks, which act not within the national economy, but as agents of the international economy and for international firms. In London (King 1990) there was a doubling of the number of foreign banks in the 1960s, and again a doubling in the 1970s. In the 1980s the numbers stabilized at about 450, but numbers employed increased in the mid-1980s from 38,000 to 72,000, to be followed by decline with the stock market crash of 1987, and then a gradual rebuild. Within this sector there is considerable specialization, including many merchant banks which deal with company finance and not with individuals, securities firms, and conglomerates, which latter include banking, securities, and currency trading as well as other functions.

Putting all the producer services together, they make a substantial although not overwhelming contribution to the employment totals. In global cities such as London, Tokyo and New York, the employment in these sectors is double the rate nationally in the countries concerned. In London (Sassen 1991), business services accounted for 10.2 per cent of all workers in the 1980s, against 5 per cent in Great

Britain in total. Growth in business services is massive, despite the 1987 crash, measured roughly by the southeast region's (London plus its outer suburbs) increase in male employment in business services plus the financial sector over 1981–94, from 10.9 per cent of employment, to 18.5 per cent.

Some of the services do not correspond to separate firms, as they are managed from within the firm. Thus, some of the legal work, the advertising, the consultations about information technology, or the firm accounting, will be conducted within the firm, often at the head offices or offices attached to them. This is indexed by the numbers of major firms that have head offices in specific cities. Among those in the top 500 biggest global firms, Tokyo is home to 34, New York to 59, London to 37, and Paris to 26. Many large cities have few such firms; Mexico City, perhaps the world's largest city today, has one of the 500, while Calcutta has none.

Within one country, the UK, of the top 500 company headquarters, 198 are located within London (*Financial Times*, 20 Jan. 1997). In Britain as elsewhere, there is some sharing of the service boom. Some banks transferred their more repetitive work to other cities, and some moved wholesale to another city, such as Lloyds Bank to Bristol. This still leaves a few cities with concentrated producer service industries.

If services are acknowledged as having dynamic potential in spatial concentrations, there are two kinds of adjustments to development theory and planning that can be made. On the "negative" side, it is of little value to bemoan the loss of a manufacturing function in some advanced countries where the process of deindustrialization has been going on for a long time. On the other hand, it is worth while including, or allowing for, services provision within any planning for development, especially, it would seem, for the advanced countries that are losing manufacturing and acquiring a dominant service role. These points are taken up in the following two sections.

Lamenting industrial decline: the deindustrialization problem

Taking first the point mentioned of worrying over lost manufacturing, a whole literature grew up in the 1980s describing and analyzing the problem of deindustrialization (Martin & Rowthorn 1986, Townsend 1993). An example is the process in the northeast of England (Hudson 1986), where unemployment rates rose from about 2 per cent in the 1950s, to 18.3 per cent in 1984. Manufacturing employment, centring around iron and steel, reduced by about 130,000 jobs over 1956–82, and coalmining by 120,000. Government policy is heavily indicted by Hudson for the decline of regional policy and support for the key industries. We may make two comments. The first is that many of the problems of unemployment were created, in the first place, by placing key industries in government hands as nationalized industries. For several decades after the Second World War, these industries were supported and their product bought by the state at unrealistic prices, because of their nationalized status. When concerns for the competitiveness of the whole British economy came to the fore in 1979, it was inevitable that a rapid cutting-back of uneconomic industries was to take place.

In addition, the industrial structure in the northeast of England was itself traditional, with few of the new growth industries. Instead, industries such as iron and steel were central, feeding heavy engineering firms producing ships and other heavy equipment. These are industries that have been effectively exported to some of the developing countries. Such industries could not easily survive without continuous support. Shipbuilding, for example, has continued its decline, with the last big firm, Swan Hunters on Tyneside, closing in 1994, although a new unit has been formed with a residual workforce. Steelmaking has been retained at Redcar, but the inland works at Consett have been closed, and employment drastically reduced overall.

There has in fact been a partial transformation whereby new employment has been created in the region, but this is not taken up by Hudson. In professional services, banking and finance there has been a doubling of employment, and this must be regarded as a positive move. Professional services existing in 1982 employed 180,000 people, double the level of 1956. More recently, the growth in automobile manufacturing at Washington, south of Newcastle, and at Sunderland, is Japanese in origin but attracts some local build-up in the supply of components for the industry. In the case of Nissan at Sunderland, opened in 1986, it is estimated that there are over 4000 jobs in the firm directly, and 6000 jobs created elsewhere in the manufacture of parts. Whereas the industrial wasteland of the northeast had only four motor parts manufacturers in 1986, it has 29 in 1996. This is a new economic base for the industry, as 70 per cent of production is exported.

Services as a positive force

Rather than manufacturing, services growth can be a basis for development in many regions. One of the largest nineteenth century traditional industrial regions in the UK is that of Glasgow and Strathclyde, where there was the same mix of coal, steel and heavy engineering as in the Tyne–Tees complex in northeast England. In this region there has been an effective shift, so that services have become lead activities. As Table 5.2 shows, there was an overall shift in Scotland towards services over the period 1971–91, but the shift has been proportionally much greater in Strathclyde, the Glasgow metropolitan region.

An example of the change is the financial services sector, now a major exporter for the region. Educational services may also be viewed as exporting, with four

Table 5.2 Employment by sector and year in Strathclyde and Scotland (in thousands).

Region	Sector	1971	1991	1993
Strathclyde				
	Manufacturing:	388.8	172.9	142.0
	Metals & engineering	219.1	87.1	
	Other manufacturing	67.2	75.8	
	Services	250.1	604.5	614.3
	Total employment	1009.6	850.7	824.8
Scotland				
	Total employment	2002.5	2003.8	1972.3
	Services	1049.9	1405.8	1451.7

Source: Scottish Abstract of Statistics, Scottish Office (1973, 1995).

universities around the city covering a range of technical and professional types of education. Some of these services link in to the growing computer and electronics industry in Silicon Glen, the corridor between Glasgow and Edinburgh. Such a development should not be exaggerated; educational services are indeed supplied all around Britain, and Glasgow is not the most prestigious provider. Financial services of a high order are still provided more by Edinburgh than Glasgow, and Silicon Glen is a region more for electronic goods assembly rather than research, the design of software, or company control and administration (Turok 1993). Over 50 per cent of the employment in the electronics industries, by 1990, was in foreign-owned firms, which normally had their central administration and research functions in other countries. The point to be made is rather that a manufacturing past does not confine a city or a region to a manufacturing future.

However, there is no inevitability that manufacturing industry will find a replacement in services. At the national level, there is a gradual move from manufacturing to services, but at the regional level, services tend to grow in complementary fashion to new manufacturing, and manufacturing decline is often accompanied by service decline. We may exemplify this from another region in Britain, the Greater London region. Here there is rapid growth of economic activity in the outer ring around London's west and north sides. The ring, or as Hall et al. (1987) termed it, the "Western Crescent", extends from Hampshire through Berkshire into Buckinghamshire, Hertfordshire and Cambridgeshire. This high-tech zone may today be added to by bringing in the M4 corridor from London to Bristol, which has attracted industry and services to a necklace of small towns along its route, such as Swindon, Bath, Chippenham, and Bristol itself. A key feature in all of this is the business services that are characteristic of modern industries – management consultancy, accountancy, market research, marketing, advertising, and computer services (Marshall et al. 1988). One kind of analysis of this Western Crescent is that it represents counter-urbanization, people fleeing the negative aspects of the modern urban environment. This holds much truth, since both light industry and business services are relatively footloose, and can choose locations where people are also happy to be living. Industry follows people, rather than people following industry.

Outside Britain, a similar pattern may be seen in France (Gaudemar & Prud'homme 1991). These writers divide France into three regions: the northeast, leaving out Alsace but with the Paris region, all of which have had heavy industrial decline; the east centre, a region including the Massif Central and the Alps, with average losses of industry; and the south and west, with only slight loss of industrial employment. In these regions, industrial change has correlated positively with service change; those areas with good industrial employment have attracted more of the service employment. In the northeast, it is thought that the large firms there brought down services with them, since these latter were ancillary services, intimately dependent on the big industries (coal, steel, heavy engineering, chemicals). In the south and west, the industries were smaller, often agro-industry, and the services were not closely related to any specific firm. Such services as tourism grew rapidly. In the south there has also been a "Sunbelt" phenomenon, with the more attractive environment bringing in small industries and services that were footloose, and depended on attracting highly trained and professional staff.

Policy issues: decentralized policy-making and city marketing

From the point of view of those planning regional policy, the stories just told are not particularly happy. As in the discussion of flexible industry patterns, most of the growth of the service sector, especially that part (business services) connected with high-tech industry, has been unaided, and not a function of regional policies. In general, neither the activities (services and high-tech industry) nor the regions concerned have been the object of policies to help them. An exception given, the Strathclyde area, has indeed been a regional development area since the 1960s, and has received monies to encourage all kinds of economic growth. But its results should not be overestimated. Rather than endogenous growth, most of the firms are incomers (Turok 1993), and the inward investment has remained controlled by outside firms.

Does all this mean that there is no place for policies to help regions in the new service economy? Hall's conclusion (1991) was that regional policy is dead, and that the remnant policies, such as enterprise zones, are points on the map, rather than extensive regions, so that policy-makers have given up on the idea that a whole region could be turned around. But there is a new kind of regional policy emerging, organized from below rather than from above. Decline of formal central government policy means a decentralization of policy to the cities and regions themselves. Their activity centres not on financial aids to industry, but on city marketing, to promote or create a favourable image for incoming and resident firms.

City marketing

Promotion of cities as places for industries to move into, on the basis of the city's infrastructure and support from the municipal authorities, was practised earliest in the USA and was prominent in the 1950s and 1960s. In Europe the idea has become popular in the 1980s and 1990s, but in a more extended form, including the promotion of the social environment and quality of life generally, with such items as the availability of city parks, lack of urban pollution, and a strong educational programme, as attractors both to endogenous firms and incomers.

This is not simply a local-scale regional programme, but constitutes a change from older concerns for welfare and help to consumers to a supply-side policy, helping to raise production and productivity through attention to the environment. It also focuses, in all recent cases, on the partnership potential between the city authorities and the private sector. New entrepreneurship is encouraged by support from the city, but not commonly or most importantly in the form of financial incentives, but in advice, consultancy, and information provision for the young firms. City marketing fits closely into the new industrial mould because the new firms' location policies, as suggested above, are based on people rather than materials, transport costs, and the like.

Again in the case of central Scotland (Paddison 1993), the city marketing of Glasgow has included special events, such as the Garden Exhibition (1989), the City of Culture (1990), and a massive effort at physical rehabilitation of the inner city and the riverside. These efforts are useful for attraction of the high-order service industries, not only for their important executive and professional staff, but also to promote a positive image of the firm itself, associated with interesting architecture or proximity to attractive scenery. In the case of Glasgow, the main need is to change the image of the city itself. Glasgow has been associated for the previous 150 years with heavy industry, working-class society, and the conflicts this economy generates between workers and factory owners, as well as with an undesirable physical environment of air, land and water pollution. This image has not wholly changed, despite the recent events and restructuring.

This kind of policy is, of course, not a real regional policy, and the degree to which the surrounding Strathclyde region benefits from it is uncertain. Such effects are likely to be seen only over a long period, in any case, during which perceptions of the city change.

Conclusions

This chapter has shown the high and rising importance of service industries in general, especially in the more advanced nations of the world. These must be seen not just as "services" to industry or to the population at large, in the sense of a purely dependent kind of activity, but as independent production sectors of the economy, and in some cases important export earners for the countries concerned.

These industries are not distributed in patterns that are dependent on manufacturing industry. Instead, and again increasingly, they have distinctive spatial patterns which the geographer is advised to study. One of the most distinctive patterns is that of the global cities, located in relation to the international economy rather than any single national market.

Services have commonly been left out of any regional policy for development. This would appear to be an error in view of their lead role in some countries. On the other hand, their locational preferences are difficult to plan *per se*, since these are industries that do not necessarily respond to classical location factors such as the presence of materials, power, or major transport lines. Some of the recent developments in service industries have depended rather on the spread of factors that can be given the encompassing name of "quality of life". The attraction of a particular location becomes one for the workers in the industry, rather than for the machines and the process involved. As a general rule, this might be expected to move major service concentrations away from traditional industrial sites, in the case of the older industrial countries, since these became disfigured by mining and heavy industry and do not present attractive landscapes. There is thus a shift, still going on, from snowbelt to sunbelt, from north to south, within the USA and Europe, towards new locations. The shift is also social, away from areas of monolithic big industry, towards areas where there are concentrations of small firms and the entrepreneurship that is responsible for them.

CHAPTER 6
Developed countries:
the United Kingdom and Spain

Apart from the consideration of theory and of the general shift in the content and orientation of development, geographers are commonly at pains to ground their ideas in observation of real world examples. Despite the onset of globalization, the best units for this observation are the individual states and their component regions. As there are such great differences between the advanced countries and those that are just now becoming industrial and urban, it is worth examining cases of both kinds. Within the countries, there are two kinds of observation to be made, first of the course of development and spatial pattern, and secondly, of the policies that have been followed to influence this pattern. For the advanced countries, two European countries are chosen: the UK and Spain.

First the actual course of development in the two countries is sketched; then there is an examination of regional development policies by governments. The UK provides important information, since it was early into the field of regional development, and has tried a variety of approaches. The most recent of these approaches will be discussed later, but policies from the recent past show how the problems have evolved and how the proposed solutions have evolved with them. Spain provides a different kind of example, a country whose development can be portrayed as having been checked after an early surge in the nineteenth century, and which is now recovering lost ground, but in ways and areas quite separate from the earlier period.

The United Kingdom

Within the UK, development has not been a constant and smooth evolution from pre-industrial times. Instead the country has pushed forward in stages when there were new inventions and changes in technology, to be followed by periods of stagnation or relative inaction. This is true for the nation as a whole, but even more so for individual regions within it. These stages have been represented, for example by Prestwich & Taylor (1990), as a set of Kondratiev waves: long waves of development involving a multiplicity of interconnected changes, in technology, in social organization, and in the use of specific resources, goods and services. On

95

the upswing of each wave, there is the introduction of new technology and a decline in the price of raw materials. This upswing is followed by a downswing, which is characterized by high prices for resources and a lack of innovation in the economy.

The first of these waves is thought to have occupied the period 1780–1840, although a case could be made that movement towards it was started many decades earlier in such embryonic industrial regions as Birmingham and Manchester. The expansive or upswing phase, from 1780 to 1820, saw the introduction of new technology for textile processing, and the first of what we would recognize as modern factories. Power use moved from water power to steam, and transport, after 1820, from canals to railways. In this period, the concentration of industry began on the coalfields of Lancashire in England and Lanarkshire in Scotland, as well as in the metropolis of London. In this way, manufacturing industry changed from what it had always been, a dispersed activity in mostly rural surroundings, scattered across many parts of the country, to a concentrated activity which brought wealth and also many environmental and social problems to particular cities and urban regions. From 1820 to 1840, there was a period of high prices for farm goods and raw materials generally, and the apparent progress of industrialization was slower.

From the time of this first growth period, there was the emergence of a sectoral concentration of industry. For the textiles industry, a leader in the period, the Lancashire cotton industry, demonstrates this spatial-sectoral concentration. As Stamp & Beaver (1954) showed in their classic geography of Britain, this grew up in the eighteenth century, pushed ahead when American cotton began to be used, and over the nineteenth century evolved into a spatially structured industry. Spinning was carried out mostly in the towns around Manchester, weaving in the spread of towns from Preston in the west to Nelson Colne in the east, and finishing and marketing concentrated in Manchester itself. The many hundreds of firms in the early industry constituted a flexible structure where all sorts of special cloths could be made by individual firms to order, and large orders shared amongst many. This was a first industrial district. It might be noted, however (see also Ch. 3), that this flexible specialization would not save the industry in the long run, for it has almost disappeared in the post-war period.

In the second Kondratiev wave, from about 1840 to 1895, a further set of technological innovations were introduced, including processes for large-scale steelmaking and the many downstream applications of steel in engineering industries such as shipbuilding, as well as development of the steam engine especially for use with railways, which themselves incorporated a stream of innovations. The period came to an end with recession, a financial and business crisis, and high prices for farm goods and minerals.

Regional concentration of industry continued, and continued to favour all the major coalfields, including the Staffordshire fields linked to pottery and engineering, the South Wales area linked to iron-making and coal export, and the coalfields of Northumberland and Durham linked to shipbuilding. All these attracted the most able people of their time into the rapidly progressing industries. They also attracted many other migrants, surplus to needs in the countryside, which was in any case depressed by low grain prices after the repeal of the Corn Laws in 1849. Many farm labourers were displaced over the first half of the nineteenth century by the Enclosure Acts, which forced a restructuring of agriculture and the elimination of many smallholdings. There was thus a general shift of prosperity from rural to urban. The first-wave industries and their regions maintained themselves with further growth: in Lancashire, cottons; in South Wales, iron and other metals. In Lanarkshire, the new iron and steel industry tended to replace the older textiles industry based on linen and cotton.

Industrial change meant changes in regional fortune. Income levels in the north, which had been comparatively low in the eighteenth century, rose to near those of the south over the nineteenth century. Smout (1986) summarized the evidence for these changes at the level of the national differences in income levels per capita for England and Scotland. In 1798, Scotland earned about 68 per cent of the level in England; by 1867, Scotland's figure was 75 per cent; and in 1911, about 95 per cent. The rise of the first modern industries in the north, followed by the heavy industries in the second wave, reduced the north–south gap to a very low level.

While the first two Kondratiev waves favoured the coalfields, especially coastal ones, the second pair of Kondratievs moved the regional focus away to new regions with different backgrounds. Because of electrical power being increasingly available, new locations were possible in the countryside and in new urban sites. The third wave, from 1895 to 1940, saw as growth industries many that were services or linked to services rather than manufacturing. Entertainments, road transport, hotel and restaurant services, printing and publishing all grew (Prestwich & Taylor 1990: 28–9). Their optimal location was in and around London, where the highest income levels were already found, and where information links to the world were best. At the same time, the old industries, particularly in the second, downswing part of the cycle, created problems of unemployment as production declined. Regions strong in textiles, in iron and steel manufacture, in heavy engineering such as shipbuilding and railway rolling stock, and coalmining, became concentrations of low purchasing power, where the whole economy and not just the main industry moved into recession, sparking the first powerful regional policies (see below).

Growth industries in this period included chemicals, the car industry, and some new consumer industries such as processed food manufacturing, but these tended to favour the new regions in the Midlands (car manufacture and light engineering) and outer London.

A fourth wave is identified from 1940 to the present. It involves a further cluster of innovations, in this case significantly not British ones but often coming from abroad, again changing regional orientation. Continuing decline in the old heavy industries, iron and steel plus heavy engineering, and in textiles, means a continuing problem in the older industrial areas.

Agricultural decline, in terms of employment, puts a further strain on rural areas, counterbalanced by new migration into the countryside, called counter-urbanization.

Part of the new wave is concerned with information technology, including telephones and computers, and the almost total replacement of all land transport systems by the road-based automobile and lorry. This is linked with the rise of services in general, which has overtaken all forms of manufacturing so that the whole country's economy is becoming based on services rather than manu-facturing. Again, in regional terms, this favours London and the southeast of England, adjacent to the continent and to Europe's central regions. For some, this latter phase of development associated with electronics, computers, and computer applications, constitutes yet another (fifth) Kondratiev wave.

It is certainly true that over the last two decades there has been a new geography beginning to appear, which is no longer exactly London-based, but in a "fertile crescent" around the capital, from East Anglia round to Dorset. Its centres are the main motorway axes, including the M4 from London to Bristol and extending into South Wales; and the M3 from London to Southampton and Portsmouth. These are the rapid-access routes important to industries that are knowledge-based and linked to the international economy. The region of concentration is highly dispersed, because of the information technology used, and because this allows factors such as the quality of life to be invoked by the workers in the new industries. They choose to live, not in the city, but in pleasant small-town surroundings, within reasonable commuting distance from the city.

The international economy argument may be extended. In the last 20 years, more than previously, if not as a totally new phenomenon, there has been the emergence of London as a global city. This term implies a function beyond that of acting as an administrative centre for the UK, or as a manufacturing centre, and instead constituting a base for activities run by MNCS: in many cases a home base, in others a base for European operations (Friedmann 1986, King 1990). Globalization does not mean simply the extension of trade and other economic links to all parts of the world. Rather, it refers to the reorganization of activities by the large MNCS, on a global basis, without any real home base or special concern for a home country. Such firms have a hierarchical organization with bases at different levels in many countries. They are also often vertically structured, with suppliers of materials and power sources, manufacturers, and assemblers for final markets, all as part of the main enterprise. Because of this, they occupy major urban centres around the world and bring large revenues to these centres through their high-paid executives and through the growth of business services such as accountants, financial services and lawyers to handle their regional problems. Globalization adds to the regional imbalances already present, in the British case, by putting more pressure on the southeast of England.

At times the regional differentiation resulting from this whole historical development process is portrayed in terms of its results, as north versus south (Smith 1989, Townsend 1993). What we emphasize from the brief survey above is that the present problem is of long emergence, and subject to a number of changes *en route* to its present status. In the nineteenth century, the north almost closed the income gap between it and the south. In the present century, the gap has reopened with the change in technology of the new waves. As a result of a long period of change, there is a north–south difference, for example in the service orientation of the south versus the industrial focus in the north, in higher unemployment in the north, and in a stronger effect of the deindustrialization trends of the 1970s and 1980s in the north. In addition, what has been attracted back into the north is a branch plant economy, as many MNCs have invested in plants producing a particular component or simply assembly plants, in neither case involving high skills or any research and development work.

Writers on the north–south divide have made a critique in terms of the government's role in boosting the south, including investment in rail links to the continent, and Ministry of Defence contracts placed with southern firms (Town and Country Planning Association 1987). This kind of critque would seem peripheral compared with the long historical process.

The north versus south contrasts may have been overemphasized, however, if we examine data on personal disposable incomes (from *Regional Trends*). The range of differences between the highest paid region of Britain, Greater London, and the lowest, Northern Ireland, changed from £1600 to £1110 in 1975, to £9348 versus £7189 (for Wales, now the lowest) in 1993 (Table 6.1). The percentage difference, using the lowest as the base, declined from 44 per cent to 30 per cent. Using a coefficient of variation weighted by population, following the method

Table 6.1 Regional differences in personal disposable income in Britain in 1975 and 1993.

Region	1975		1993	
	Population (millions)	Personal income (£)	Population (millions)	Personal income (£)
UK total	55.44	1331	58.19	7942
North	3.12	1234	3.10	7246
Yorks. and Humberside	4.90	1241	5.01	7437
E. Midlands	3.73	1278	4.08	7477
E. Anglia	1.79	1252	2.08	8055
Greater London	7.11	1600	6.93	9348
Rest of southeast	9.81	1371	10.84	8288
Southwest	4.23	1274	4.77	7967
West Midlands	5.18	1350	5.29	7622
Northwest	6.57	1297	6.41	7454
Wales	3.44	1231	2.91	7189
Scotland	5.21	1288	5.12	8065
Northern Ireland	1.54	1110	1.63	7413

Source: Regional Trends **26** (1995) Table 12.9, personal disposable income.

established by Williamson (1965), the differences decline from 0.65 to 0.60. These indicators of a reducing difference run contrary to the usual presentation of a large and growing divide between the northern and southern parts of the country.

Thus the disparities forecast by the north–south researchers are not evident. According to Martin (1986), "of all the forms of state policy, the freemarket conservatism of the Thatcher government is the most likely to exacerbate widening disparities in the British industrial landscape". On the other hand, Martin himself, in the same book (Ch. 1), noted that industrial maturity means that dynamism in Britain moves on from the nineteenth century iron and steel industries to other sectors and other regions. There is no hope of recovering the past.

British regional policy

Regional development in Britain has mostly been directed at industrial problem regions. This basic fact needs emphasizing, since other approaches might have been possible, focusing, for example, on peripheral regions, growth regions or rural regions. As a result of the focus, most efforts have been towards retaining existing industries or bringing in new ones. The history starts in the period after the First World War, when production levels declined after the huge war effort. World depression in the 1930s was already anticipated in Britain by the 1920s; the force of this depression was felt most in the industrial zones set up in the nineteenth century, where coalmining, iron and steel industry, and heavy engineering were dominant (Fig. 6.1). These regions included central Scotland, the Northumberland and Durham coalfield centring on Tyneside, the Yorkshire industrial towns, the South Wales coalfield, the south Lancashire cotton district, and Northern Ireland. In contrast to these, as already outlined, regions in the south were experiencing growth, with their chemicals industries, car manufacturing and electrical goods. In addition, the services sector was growing, again mostly in the south. Rural areas in Britain were also in depression conditions, but they represented no concentrated block of unemployed people as did the industrial zones, and the awareness of the industrial zones was heightened by the actions of unions and protest movements at the closure of mines and factories.

High levels of awareness of the problem through the media, and the political problem of strikes and protest marches, made some kind of reaction inevitable. A first attempt was through stimulating the movement of labour. From 1921 onwards, workers began to move with encouragement from firms, and in 1928 the government established an Industrial Transference Board to give formal aid to movement. About 650,000 people moved to the growth areas over 1921–37, nearly 600,000 being aided by the Board which covered movement costs and gave retraining to workers from such industries as coalmining.

This is a policy that may be characterized loosely as coming from the orthodox, right-wing stable, for it indicates simply the need to help overcome the rigidities in

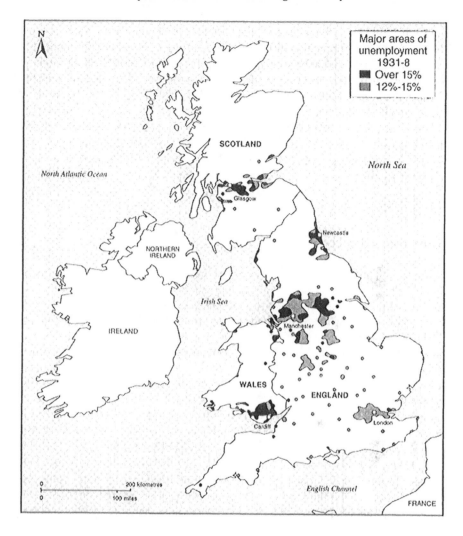

Figure 6.1 Areas of high unemployment in Britain in 1931–8, corresponding well with the regions of nineteenth century industrialization of the first two Kondratiev waves.

the labour market by moving people to the jobs. This policy was probably effective, but not popular because of the social disturbances it created. It was not welcomed by the firms and families in the regions of origin, which, in a process of selective migration, saw the most able of their members leaving for the south. From 1934 it was succeeded by the Special Areas Act, which sought to alleviate the unemployment problem in four Special Areas, central Scotland, the northeast, west Cumberland, and South Wales, with a separate arrangement for Northern Ireland. Within these areas the Board used the device of industrial estates, providing the infrastructure for new factories on estates at Cardiff, Gateshead and

101

Glasgow, and also building the factories for the firms to start up in, as well as offering loans to firms setting up in the Special Areas. Outside industry, money was spent on water and sewerage improvements, upgrading and landscaping of derelict sites, and some urban improvements such as the creation of parks. In the countryside, smallholdings with their own cottages and allotments were set up to help poor rural people.

Actual policy in the between-war period could still be regarded as an orthodox policy, since it sought to help industries into movements where the movement costs were large. In the event, the policies had little real effect if measured in terms of industrial growth between the wars. In the southeast in the 1930s, Greater London received 45 per cent of the employment in new factories, and the Special Areas less than 5 per cent.

Post-war Britain

Towards the end of the inter-war period, in 1938, the Barlow Commission (its full name being the Royal Commission on the Geographical Distribution of the Industrial Population) was appointed to review industrial change, and its report in 1940 set the scene for the post-war era by arguing that intervention was needed both to attract industry into the declining regions, and to restrict the growth of the southeast. This set in place the concept of the "carrot and stick" approach after 1945, with encouragement to the weaker regions, but restriction on the more attractive. Barlow argued for an interventionist approach because markets for labour and capital were not self-balancing, but leading to long-term and serious imbalance. Much more powerful regional policies were enacted in the post-war years, from the early 1950s on. On the basis of unemployment rates, a set of regions was defined that again included the old industrial regions dominated by mining and nineteenth-century industries, which found themselves dropping back quickly into depression. War industries were run down and the leading edge of technology passed them by, moving on to chemicals and electrical goods, so that these areas found their industries had competitors in overseas countries (e.g. in steel, shipbuilding, textiles), taking away their markets.

The measures used included ones both of overt regional policy and indirect policy. Overt policy included the "stick" policies to move manufacturing industry out of the southeast. Firms were prohibited from building large factories (over 10,000 sq. ft. initially, then over 5000 sq. ft, and in 1965 finally 1000 sq. ft) without a special certificate in the southeast and Midlands, while they were given considerable inducements in tax concessions, advance factories and low-cost credits to move to the Development Areas.

On the "carrot" side, the pre-war measures of industrial estates, infrastructure building and loans to new factory projects were continued, in what were now called Development Areas rather than the old Special Areas. The programmes became more complex over time, with one addition being a subsidy to employment, the Regional Employment Premium (1967), and another being the definition of Special Development Areas (1967): those with acute employment problems, which were given extra financial aid. Intermediate Areas (1970) were also designated, based on their having characteristics such as poor physical infrastructure and high out-migration as well as moderately high unemployment.

The above account is an outline of what is usually discussed as regional policy in Britain. This account misses a large, probably more significant segment of policy affecting the regions where much larger amounts of capital were invested over the years than in direct regional policy. This was through state ownership of some of the heavy industries dating to the first two Kondratiev cycles. In the period from 1945 to 1950, several of these industries were taken over by the state through the Labour government of the time. They included iron and steel (nationalized briefly from 1949, then again from 1967), coalmining from 1947, gas from 1949, electricity from 1948 (in Scotland from 1943), and the major transport systems, including railways, from 1947, as well as a variety of other industries.

It is important to note how these nationalizations, and the subsequent support given to the respective industries, formed a regional policy. It is precisely because they were industries located in the north and west, in association with coalfields, and industries from the first two Kondratiev waves, that a regional pattern of support emerges. Thus nationalization of coalmining, for example, became very much a support system for the coalfield regions of South Wales, Yorkshire and Lancashire, the Scottish lowlands, and the Northumberland and Durham region. Government policy was to maintain the coal pits open for as long as possible, through subsidy, and to find markets for the coal in the power-stations which were also government owned. Iron, steel and heavy engineering, the "commanding heights of the economy", which prevailing socialist thought regarded as necessary to come under government tutelage, were also the subject of massive subsidy and support through the provision of a protected market, in maintenance of the nationalized railway stock and track, or in the building of ships for the Royal Navy. These industries began to be denationalized or privatized in the 1980s, in a process that continued up to 1996.

A case may be made that the huge efforts made to retain these industries were in fact a waste of national government budgets. Coal was found to be uneconomic to mine in Britain, compared with imported products and alternative power sources. Heavy engineering could not meet the competition of many foreign countries, including some that supported their own industries in the Far East and in Latin America, while enjoying lower labour and material costs. All these industries were apparently ones that Britain would have needed to leave in the face of international competition. There is a comparison that can be made here with some LDCs, such as those in most of Latin America, which have devoted attention to setting up new industries in remote interior regions, at the expense of central

103

development, for populist and geopolitical reasons. In Britain, it is possible to claim that the old industrial regions were a diversion which caused a decline in international economic standing in the whole period from 1945 to 1980.

Another arm of indirect policy was through the welfare services, unemployment and pensions benefits, health, education and housing. Since in the poor regions there were high levels of unemployment, low levels of home ownership, and weaker educational provision, these regions would benefit from state welfare. Robinson has shown convincingly that the middle classes made most use, per capita, of educational facilities and the National Health Service in the 1970s (Le Grand 1982); but without national welfare, the provision of housing, education and other welfare would have been slight in the poorer regions, and disparities between regions might have been still higher.

Comprehensive economic planning regions

It is worth mentioning a different kind of regional definition and development body. From 1976, Scottish and Welsh Development Agencies were set up. The differences from the usual regional aid programmes were in the autonomy of these bodies to administer their grants from central government, not earmarked for specific projects or controlled from London, so that in each case, a more varied and complex system of aid came into being. These agencies have survived to the present day and will be the subject of later comment.

Another form of regionalization for comprehensive economic planning was undertaken from 1964. In that year, a Department of Economic Planning was established, and prepared a National Plan to be administered through a whole set of strategic plans for each of 11 major regions. These were to be agreed by composite bodies representing central government, local government councils, and a regional planning council. However, the National Plan was itself abandoned after two years, and the Department of Economic Planning from 1969. Strategic plans were written for several regions, but they lost their impetus with the loss of a national framework (Damesick 1987). After the election of the Conservative government from 1979, there was a winding down of the formal regional policy, leaving the Development Areas in place but with smaller attractions on offer. There are traditional credits and tax reductions available, but on a limited scale and most strongly in the Enterprise Zones set up in the 1980s. In addition, indirect policy through the operation of the nationalized industries is being gradually removed through privatization of the relevant industries.

A critique

The whole period up to about 1980 may be regarded as one of regional policy based loosely on a welfare analysis; this policy addressed the levels of consumption and welfare in disadvantaged regions, more than production and improvements in productivity. There were aids to private manufacturing industry, but of greater regional effect were the monies paid in welfare to the regions, and in support for the nationalized industries. Because of the location of these industries, there was a strong regional dimension to aid.

If the state took over some industries, it still did not use the official planning machinery to undertake strong planning. In Britain, as in most of the developed world, planning was "indicative", guiding developments rather than specifying them. Economic planning could thus never be very powerful. What was carried out carefully and successfully was physical planning, meaning the planning of land use, containing urban growth, designating conservation areas and green belts, and creating "new towns". This continues to be a strong element, but is not accompanied by strong economic measures to ensure that firms and types of production are located at specific points.

In retrospect it is possible to see that much of the policy, overt or covert, was a failure. Overt policy tried to attract firms to distant, often unwelcoming environments in industrial districts with much derelict land. The firm's commitment to such areas was understandably not high, and many factories were branch plants of companies with main factories in the south. The policy was not sufficiently detailed, either, to bring together structurally related clusters of firms, so that each new factory was isolated in the development zone. Attractions were set at the same level for biscuit factories and for car tyre manufacturers; thus a major industrial estate at Drumchapel, west Glasgow, from 1960, has been abandoned for several years, both the biscuit factory and the tyre manufacturer, Firestone, using the site with no reference to each other or to other local firms. In the same period, the Rootes organization (later absorbed into Chrysler), based in the Coventry area making cars, was induced to establish a new plant at Linwood, near Paisley in Scotland, with the idea of forming a new kind of industrial complex in the west of Scotland. However, the factory failed to attract new component manufacturers, and components continued to be supplied from the south. In the end, closure of the factory was inevitable. At the level of the nationalized industries, the policy was also regionally flawed. British Steel had a major plant at Ravenscraig, east of Glasgow, but it was closed in 1993 in an economy drive by the company. This huge integrated plant had been located in Scotland in 1954, as a sop to Scottish complaints that Wales was being favoured, but contrary to economic dictates. In the end, the factory closed, and the period of 40 years could be seen as merely delaying the restructuring of an industrial region. For a country like Britain, with a consumption of around 20 million tons of steel annually, the most efficient size of mill was about 10 million tons, and thus only two mills were really needed, with a few small ones for special steels. Ravenscraig would always be in excess of requirements.

In general, as the state industries were those having problems (i.e. the older industries inherited from the nineteenth century), the help the government could give was towards retaining old industry, rather than bringing in new industry.

Within the advanced countries, this has been a poor strategy, as growth has come to those regions and countries concentrating on new technologies and products.

The spatial structure

Apart from the results, it is possible to criticize the whole spatial apparatus of regional aid to development in Britain. For the planning to be effective, a regional structure of economic planning regions might have been used, as was begun in 1964, dividing the whole country into units within which specific plans might be set up. Such regions could be devised so as to maximize both their homogeneity (e.g. coalfield regions, textile regions) and their functional identity, focusing on one of the large regional cities. As an alternative for regional planning, the counties have never had a strong economic function. Instead of a strategic planning system, there came into being a set of regions, varying in geography over time, defined on the basis of a single set of statistics: unemployment.

Perhaps an even more fundamental critique is that the definition of regions for aid was not a function of positive aims, but a fire-fighting approach, seeking to help every region that encountered difficulty. This is attacking the symptoms rather than the causes. Only in a few noteworthy exceptions were there moves to new industrial areas. The best-known of these is the "new towns", which were created as industrial sites as well as residential areas for overflow from the large cities.

It is hard to avoid the conclusion that much of the money spent must have been incorrectly allocated to industries and areas with little hope, and that the forces for change and development were simply hindered by the programme. The lesson that can be gained from a study of economic history of British regions is that a series of massive changes has occurred, and that while some regions can convert so as to have important roles in more than one long wave of development, many are important only during a single period when they enjoy a favourable combination of physical and human resources.

Spain

This country belongs to Europe and, on some historical indexes, to the older industrialized, developed group of countries like the UK. On the other hand, it has also been regarded as part of the semi-periphery in studies of world systems (Knox & Agnew 1989). Under this analysis, rather than part of the core, it is on a Mediterranean fringe, with the alternative possibilities of moving up into the core, or down into the periphery. It offers some parallels with Britain, but also some significant differences which are treated in the following summary of development. One of the differences has been in the strength of regional feeling. Spain was united in the fifteenth century with the unification of Castille and Aragon, and the expulsion of the Moors in Andalucia. But it retains a strong regional element in its cultures, partially indicated by the continued existence of regional languages, Catalan and Basque being the strongest, although Galician is also receiving modern attention in the resurgence of regionalism since 1976, after Franco's long period of centralism. Modern industrial-based economic development in Spain came in the late nineteenth century, was checked by the Civil War and Franco's 40-year reign, and has moved rapidly ahead since 1976. As distinct from Britain, recent development has been heavily dependent on special resources, Spain's fine climate attracting tourists and allowing specialist agriculture to produce for the rest of Europe.

Regional economic differences in the level of incomes and generally the standard of living are greater in Spain than in Britain. Compared with differences in per capita income of about 1.5 to 1.0 between the highest and lowest regions of the UK, the differences between the richest and poorest provinces of Spain, as late as 1989, were of the order of 2.3 to 1.0 (Banco Bilbao Vizcaya 1991) (Fig. 6.2). Powerful currents of migration from south to north, mostly from the region of Andalucia to the cities of Madrid and Barcelona, and intense development in a

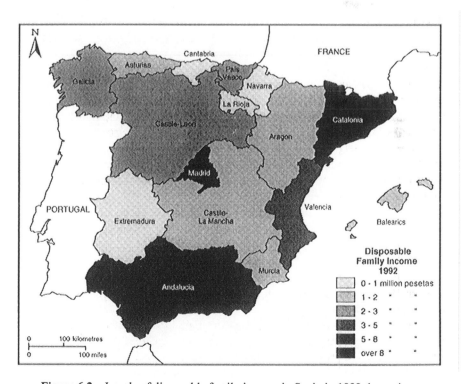

Figure 6.2   Levels of disposable family income in Spain in 1992, by region.

few regions, characterize the Spanish record of the twentieth century. In 1970, Madrid and Barcelona provinces each had about 1.8 million lifetime migrants, of which the great majority came from southern provinces (Barbancho 1979). It is this movement, rather than regional policy, in a country whose total 1970 population was only about 30 million, that has been responsible for balancing between the more developed and less developed regions. As recently as 1955, the per capita income differences had a range of 4.4 to 1.0, and they declined to their present level entirely between that date and 1975.

Modernization and industrialization came later to Spain than to Britain, and this is a central difference between them, affecting not only the stage of development but also the path followed. Spain had a traditional agrarian society and economy up to the beginning of the twentieth century. Some aspects of agrarian reform were attempted in the nineteenth century with the dispossession of Church properties, but this only led to the shift of large estates from clerical to lay hands. The counterpart of the large estates was many landless labourers and excessively small farms alongside the big underworked estates. By 1900, only 16 per cent of the working population was in the manufacturing sector of the economy, and most people were in farming families. Even by 1930, only 26.5 per cent were in manufacturing, including construction. What manufacturing had developed was only that common in the early phases of industrialization. In 1925, clothing, food and drink accounted for 58 per cent of the manufacturing employment, and was still 39 per cent in 1950. Up to 1950, the only manufacturing in many regions of Spain was craft industry with some modernization, such as the fish processing and packing and leather industries of Galicia.

Spain's maximum rate of employment in manufacturing and construction was in 1985, and the dominant sector is now services, with 52.8 per cent of employment in 1992, as opposed to 31 per cent in industry and only 9.5 per cent in agriculture. Part of the service growth is of tourism; contrary to most policy initiatives for national and regional development, this service sector has proved to be the lead sector in Spain. In an older industrialized region, Catalonia, during the 1980s, it provided 10 per cent of GDP. In Spain as a whole, it provided over 20 per cent of GDP.

If we try to compare the Kondratiev wave model of British development with that of Spain, there are some important differences to be noted. First, the different technological innovations come later and from the outside, as exogenous rather than endogenous processes. Secondly, the timing of these waves is conditioned not so much by the innovations themselves, but by the degree of linkage to the outside world, and by the actions of government in promoting or hindering economic change.

A first industrial revolution may be identified in Catalonia from 1830 to 1860, when the catalyst was the cotton and later woollen textile industry (Carreras 1990). A massive and modern industry built up in the industrial satellite towns around Barcelona, Tarrasa and Sabadell, using large amounts of coal and steam power like Lancashire. The industry had grown first in the eighteenth century, but was checked by the Napoleonic Wars, by the loss of the American colonies, and by civil wars in Spain, so that it only settled again after a long pause. By 1856, the Barcelona urban region had become an industrial district of some size, and Catalonia as a whole was dominated by this industry, which accounted for 61 per cent of manufacturing in the region. By the end of the century there was further growth and diversification into other textile industries, including silk and linen.

A second Kondratiev wave, occurring roughly between 1880 and 1914, was centred on Spain's north coast, Cantabria (Nadal & Carreras 1990). Part of it was based on the ria of Bilbao itself, where there had long been an iron ore export industry, which allowed the accumulation of some capital amongst the mining and merchant firms responsible for shipping the ore. Steel manufacture was followed with a shipbuilding and heavy engineering industry based on the large iron and steel mills set up around 1880. In Santander, further west, the dominant industries were forges and presses making engineering products. For Asturias at the western fringe, coalmining was predominant, but there were also local iron and steelworks.

To the east of Bilbao in Guipuzcoa, a highly diversified industrial complex came into being in this late nineteenth century period. This was based on a variety of industries, including an important paper industry at Tolosa, and textile mills along the small rivers of this mountainous region. Many of these industries were set up by French and Catalan entrepreneurs. Beer, chocolate, matches and soap were other small industries, and in the interior both of Guipuzcoa and Vizcaya, there were many craft foundries and skilled engineering industries such as the manufacture of small arms, of domestic stoves, of cutlery and of locks, which were integrated into the steelmaking industries growing on the coast.

In contrast to Barcelona's development of textiles, Cantabrian industry was subject to government intervention and protection from foreign competition by physical limits on imports and by high tariff barriers, especially after the advent of Franco. Franco's Spain, as set up after the Civil War, was based on the principle of self-sufficiency, and in the early years, isolation and independence from the rest of Europe. This stance was only gradually relaxed after the Second World War. Barriers to market competition were maintained in place up to 1958, when the system became untenable and Spain was forced to adopt a more open economic policy, beginning its movement from a protected to a free-market economy.

In the period since 1958, the opening of the economy has meant an effective fusing together of later innovations in a single complex mass of changes in national and regional economy. No longer is it possible to talk of a Kondratiev wave led by a particular innovation or set of innovations. Instead, there are changes in government and the amount of restrictions or controls it exerts, allied to changes in national economic orientation and outside tastes and preferences, As a result, Spain has become host to a variety of new industries and economic activities, in new regions of development, hitherto considered economic back-waters. Sectorally, one change is towards the tertiary sector, and particularly towards tourism in the new Spain, which favours the Mediterranean coasts. Another change is towards new light industries, in small and medium-sized firms, and away from the old large industries such as those of the Cantabrian coast. These have declined as a result of increasing international competition, which has been felt increasingly as the national economy has opened, first timidly in the 1960s, then more strongly in the 1970s and 1980s. A further change worth noting is that towards incorporation of Spain into European markets for food products. This has driven the expansion of fruit and vegetable growing for export markets, replacing traditional crops which were those needed to feed the protected domestic market of Spain itself. A final set of changes is the attractiveness that a more open Spain represented for international firms after 1960. This brought in labour-intensive manufacturing such as the car manufacturers, to cities such as Valencia, Barcelona, Madrid and Zaragoza, where they had a combination of relatively cheap labour and good access to the European market.

The regions that emerge as those of greatest economic growth are the whole Mediterranean coast, from Gerona north of Barcelona, south through Valencia and Murcia, to the Andalucian Costa del Sol. There is also accelerated growth in the Ebro axis from Barcelona towards the northwest (Garcia Delgado 1991, Chs 18–20). In these regions there is a different structure of firms from that in the older industrial zones (Vasquez Barquero 1990). Many small firms with fewer than ten employees are found, and many new firms have been created in the intermediate-size towns and cities of the regions. Much of the growth is related to new powers given to local governments, which are acting often in concert with private enterprise to create industrial parks and provide infrastructure for new firms (Vasquez Barquero 1990).

This collection of massive changes, especially since 1960, has had only modest effects on the patterns of interregional disparity in incomes and in standards of living. Because of the decline of some more prosperous regions like that of Bilbao, at the same time as the rise of some poorer ones as on the southern Mediterranean coast, the income levels are converging over time. There have been measures of the standard of living based on a variety of indicators (Barke 1989, Barke and Park 1994) that show that these have also converged, although somewhat later, in the 1980s. On such social indicators as the number of cinemas, the number of suicides, and the number of telephone calls per thousand population, and welfare provision such as schools and doctors, the trends were towards interregional equality in the 1980s. On the other hand, summing the results, there remains high diversity because of the extreme values of social disorder in some large cities such as Barcelona and Madrid, where there are high rates of crime and vehicle accidents. These results are to be anticipated, in a country undergoing a rapid urbanization and industrialization trend.

The explanation in terms of Kondratiev waves does not entirely fit this evolution of Spanish spatial economy. While early stages seem to correspond to the Kondratiev model, the late twentieth century changes have come as a mixture of developments resulting, not from technological innovations, but from new markets and new evaluations of resources.

One alternative way to view recent Spanish development patterns is indeed in terms of a new resource orientation. For the USA, a general tendency has been observed since the 1960s for industries to decline in the northeastern states of older industrialization, and for new industries to rise that are associated either with the resources of oil and irrigated farming, or with quality-of-life conditions for the workers and especially the executives of firms. This general trend has been described as the rise of the "Sunbelt" as opposed to the "Snowbelt" in the north, or more derogatorily, the "Rustbelt", in reference to outdated manufacturing industries (Sternlieb & Hughes 1978).

An argument can be made that there is a new sunbelt in the south of Europe, contrasting with the snowbelt of the northern countries and regions. On this analysis, the southern regions have attracted new industries and other activities because they provide a more attractive environment for the executives and other workers. Whereas older industries were confined by their need for mining products, transport and access to markets, the new ones are footloose and can choose new regions where people enjoy living. Spain, particularly coastal Spain, provides a pleasant physical environment with a good climate, scenery, access to water and mountains, while the local urban environment is also pleasant, with many small cities offering good facilities for education, health and housing.

Development of intensive irrigated agriculture has certainly benefited from the fine Mediterranean winter climate, providing export crops to western Europe. The tourism sector, including hotels, restaurants and all leisure facilities, has equally benefited from the positive environmental features (Morris 1992b). There is a need, however, to extend this environmental or resource model of regional development. Much of the new development has been in the small industries of the Mediterranean coast, which have been successful in growing to feed both domestic and export markets. The manufacturing industries are in such sectors as electronics (Barcelona), car manufacturing (Valencia), ceramics (Castellon), leather goods and shoes (Alicante). Many rely on flexible labour forces with moderate skills, and much family labour. In many ways they are like the firms of the Third Italy, and for comparable reasons. Having been bypassed by the earlier industrialization phases, they held on to small family businesses and to the idea of entrepreneurship, which has thereafter benefited them when the opportunity for growth appeared (Naylon 1992).

Feeding all the growth sectors has been a major migration process in Spain, which has seen at least 5.7 million people enter into interregional migration over the 1962–76 period (Barke & Park 1994). Much of the migration was from Andalucia to Madrid and Barcelona, a movement in classical form to combine northern capital with southern cheap labour, giving Spain a competitive advantage in many industries with moderate skill requirements.

A third strand of explanation, beyond that of the sunbelt and flexible industries, concerns the role of decentralization, which has been occurring in Spain since Franco. In the more dynamic regions of Spain, local enterprise has seen cooperation between local firms and local government since 1978, and the creation of the 17 Spanish regions (Vasquez Barquero 1986). There has been a partial dismantling of central control, and the initiative in regional development has passed from Madrid to regional centres like Barcelona or to lower levels like the municipalities.

Regional planning

Compared with that of the UK, regional planning in Spain has been of lesser significance and little impact. The equalization of interregional differences, taken to be the main aim of policy in British regional planning, was never set as a central target in Spain. If it had been, the efforts would have had to concentrate on the south, the area of greatest poverty, with a large farming population and many landless labourers. This was the object of reform attempted by the civilian government of 1932–3, which would have carried through a complete agrarian reform. This proposal so antagonized local landowners and their supporters in the military that it was a major trigger for the Civil War in Spain.

Instead, the Franco government eventually turned to a different approach based on the French model of growth poles. Twelve intermediate cities were established as poles from 1964, with their aim being the simple one of decentralizing industrial growth away from Madrid and Barcelona (Richardson 1975). The list of poles is: Burgos, Huelva, La Coruña, Seville, Valladolid, Vigo and Zaragoza, all nominated in 1964; and Granada, Cordoba, Oviedo, Logroño, and Villagarcia de Arosa, in the years from 1970 to 1972. These were not located in the poorest provinces, but mostly intermediate ones. They offered some pre-existing industrial development, and lay on important development axes either in existence or planned.

At each of the poles, a set of financial incentives was put in place to encourage new firms to establish themselves. The duration of special conditions was originally only five years, but had to be adjusted to ten to achieve any results. Tax relief on profits for the early years, subsidy on capital investments, tax remission on imports needed to set up the firm, and preference in obtaining official credit, all brought in footloose industries. This was, however, insufficient to offset general trends which meant that the pole provinces lost their share of Spanish manufacturing, falling from 19.5 per cent in 1962, to 18.8 per cent by 1975 (Mendez Gutierrez del Valle 1990). Richardson (1975) made some estimates of the effectiveness of the poles in transmitting growth to their surrounding areas, and found generally poor results. Seventy per cent of purchases by pole firms, and 80 per cent of sales, were outside the pole's own province.

More importantly, the investment made in the poles policy was insufficient to offset the macroeconomic attractions of Spain as a site for MNCs to set up in the

1960s, and the big cities were the obvious target for these, with their better transport linkages, physical infrastructure, and large labour pools. Rather than intermediate cities such as Huelva and Seville, designated as poles, Madrid and Barcelona were the recipients of major inward investment.

There have been other policies with a potential regional effect, such as the INI (Instituto Nacional de Industria), the national industrial holding company that accumulated over 70 firms which it sought to maintain as strategic industries or in need of temporary help. There has also been a modest agrarian reform policy. But in Spain memories are long, and the 1930s type of agrarian reform by expropriation, a major contributor to the Civil War, has not been attempted again. Instead, this policy has concentrated on agricultural colonization projects, mostly to extend irrigation agriculture on to new lands. But these actions have not been regionally directed and have had little impact since little state investment was put into them.

After the death of Franco, a new Constitution for Spain was formulated in 1978, which created 17 autonomous regions. The verdict on this redistribution of power resources has not yet been given, but it may be noted that the regions are of very uneven character, and their devolution has moved at different paces. The strongest units with most political voice, Catalonia and the Basque Country, have devolved further and more effectively than many smaller and less politically identified regions such as Murcia, really only a left-over from the carve-up of the provincial system. Decentralization also has had major costs, in the setting up of another layer of local government, alongside the existing municipalities, provinces and national government. Because autonomy includes substantial control over the regional economies (Barquero & Hebbert 1985), there is also room for greater divergence in both the style and level of regional development. Reconstruction of old heavy industries is high on the list of priorities for the Basque area, while new development is the aim of Andalucia. These aims may have very different consequences, and so decentralization may be regarded as having no particular equalizing effect.

Vasquez Barquero (1992), in another contribution on recent development, shows how the growth of new small industry has in fact been concentrated along the Mediterranean coast, from Gerona down to Valencia and Murcia, and in the Balearics, where the tertiary sector has been strong and where governmental influence of any kind has been quite low. Local municipal help and support to new firms is of strong and rising significance for this new kind of development thrust.

Conclusions

As commented on above, the recent regional policy in Spain has been politicized, through the measures of decentralization undertaken after the death of Franco. A national economic policy for regional development is now scarcely in existence, although there are funds for transfer of tax revenues from richer regions to poorer ones. Most of Spain gains now from European policies which are regionally differentiated. EU aid is mostly given through the so-called Structural Funds, which were reformed in 1988 and provide the main form of aid outside agricultural support. The main reform was the creation of Objective 1, which uses 70 per cent of the funds, and is intended for Regions of Lagging development. Almost all of Spain outside the northeast comes into this category. In addition, Catalonia and the Basque Country qualify for the second category, Objective 2, for Regions in Industrial Decline. Rural areas throughout Spain have funds under Objective 5b, promoting Rural Diversification.

Overall, the message given by Spanish regional development is that the country has produced only a weak regional policy. On the other hand, there has been a massive set of interregional forces for change in the latter half of the twentieth century. Some of these forces are exogenous, like tourist demand and inward investment, partly outside the control of domestic governments, and have produced a greater concentration of wealth and production in some regions. The ability of the country to influence waves of development affecting this or that region is limited because of the exogenous character of change. The balancing that has occurred in recent decades may be ascribed more to "natural movements", the migration of several million people from Andalucia to the north, and to some extent the movement of capital into the south to exploit new resources such as the climate. Both because of this exogenous nature, and because of the weak finances of the state in Spain, as can probably be said for the peripheral countries of southern and eastern Europe generally, planning regional development is a less fruitful task than in older developed countries.

CHAPTER 7

Less developed countries: Argentina, India and the former Soviet Union

There are some major differences to note between the development structure and pattern identified in Europe, and those of the less developed countries or LDCs. These differences suggest that distinctive development policies may be required in such countries, and indeed standard European-type policies have not been successful when applied to these countries. The terms "North" and "South" are sometimes used as a shorthand for the rich–poor division, but this term is not used here as it has political connotations, being the logical sequence to the "Third World", which places the poorer countries as a third bloc lying outside the First World (North America and Europe) and the Second World (the Communist bloc). North–South suggests a confrontation and an acceptance of the dependency arguments, whereas in fact the LDCs that have been able to link well to the developed countries (like the NICs) have been development successes, while others such as Cuba and for a while Tanzania, which tried to cut themselves off from the West, have a poor record.

One set of differences about LDCs relates to the course of development. For many of these countries, the concept of development is scarcely relevant – they are unable to make any progress in the material standards of living, and stand at subsistence level over long periods of time. Their income levels, at under US$1000 per capita per annum in sub-Saharan Africa, and around $2000 in Latin America, stand at a quarter or a fifth that of the developed countries of Europe, and seem to have been at that level over a long period of time. Data collected for Latin America suggest the income gaps between the regions of each country have had no obvious trend at all over the last 30 years, so that there is no convergence internally. These gaps can also be very large, much larger than the 1:1.5 level (the ratio of the poorest region's per capita income to that of the richest region) of Britain, or the 1:3 level in Spain, and reaching 1:10 in a country like Argentina. Such data are totally dependent on the units of definition, but they differ so widely as to make very likely the existence of major differences.

This may be interpreted in terms of Williamson's (1965) convergence model for interregional difference, which envisages an increase, then a gradual convergence of regional per capita incomes over time. Countries such as India and China may be regarded as furthest behind on the development trajectory, so that their differences may be expected to increase, then reduce over time, in a natural manner. This is the benign interpretation, but this is only allowable if the course of development for these countries is going to be the same as for the presently advanced countries, which seems an untenable proposition. Leads have been established in industrial innovation, infrastructure and education of the populace which present problems for newcomers. There is even a dependency, in the limited sense of reliance on outside technology, capital and trade relations, for advance to be made, although none of these factors is insuperable.

In terms of Alonso's five bell shapes in development (see Ch. 1), it appears that the fourth, population and industrial concentration in one or a few cities, and the fifth, population growth, move in exaggerated form, and will drive the rates for the other elements. Taking the fourth bell shape, in the poor countries there have been spectacular increases in the levels of social communication, with the advent of radio and television, greater literacy and availability of written materials, all occurring ahead of other changes, so that the poorest peasants in the most remote rural periphery are made aware of the levels of living in the wealthiest cities of their country. This change in information levels is little studied or used in the literature, but it must form a major difference from past centuries in Europe, when awareness of conditions in other parts of the same country, let alone other parts of the world, was slight and often inaccurate. In Europe, industrial urban development took place at the same time as an increase in awareness. Because of this greater early awareness, before any strong growth in the economy, and the "revolution of rising expectations" which it engenders amongst the poor, there is a greater movement from countryside to town. This movement, commonly known as rural–urban migration, is no doubt further aided by the improvements in transport conditions.

As to the fifth bell shape, population growth, it is the case that this has been a force of exaggerated proportions in some poorer countries. Help from the rich countries to the poor has often been first in the area of sanitation, health and medical care; this has the result that population, previously held back by natural curbs such as malnutrition, disease and war, increases rapidly. Help from countries with the Christian tradition focuses on ensuring the survival and health of babies and young children, a focus that concentrates specifically on the immediate cause of population growth. Again, the demographic transition starts before transitions in other elements of society and economy, and serves to check any emergent development.

Movement in the fourth bell shape, population concentration, does help to slow population growth, but some part of the population growth is independent of this urbanization, and constitutes a major hazard for the countries concerned, if not an ecological hazard for the whole world because of the huge drain on world resources.

Groups of countries

It is important to note some major differences between the various LDCs. A number of different attempts have been made to classify these, but a simple geographical split encompasses much of the variation (Table 7.1). If we concentrate on the former Third World, the three areas of interest are southeast and south Asia, Africa, and Latin America. These can be assembled into four major groupings of countries. The east Asian newly industrialized countries (NICS) merit a group on their own, because they have so obviously moved away from the other countries in southeast Asia.

Evidently there is some overlap between the four groupings above, but they have a clear overall distinction. Latin America is characterized by GNP per capita levels of between roughly $1000 and $7000 per annum, a modest contribution of agriculture to this GNP despite their large land areas, high levels of urbanization (over 60 per cent commonly), and moderate to high levels of population increase, although given their large land reserves, this is not too critical at the moment. Sub-Saharan Africa has annual per capita GNP levels running from about $100 to $800. Of this, a third to a half is generated through agriculture, in countries that are still dominantly rural rather than urban. Population growth rates are excessively high. In southeast Asia, there are two groups identified: the LDCs and the NICS. Of these, the LDCs are comparable to the African countries, with very low GNP, under $500 typically, and up to half coming from agriculture, with low levels of urbanization and rapid population growth. The NICS, however, have a wide range of values for GNP, for contributions from agriculture, for urbanization and for population growth. Amongst these countries, there appears to be a rapid diffusion process going on; its centre is the dynamo of Japan, which in retrospect can be identified as the first NIC, its economic development starting before the Second World War and coming to fruition in the post-war era.

Failures and successes

Some idea of the nature of the development process and policies in the LDCs can be obtained from examples. Let us take first the examples of Argentina and India.

Argentina

This country may be used to exemplify a "Latin American type", although there are obvious differences from country to country within that world region. It is of very large areal size, with 2.8 million km^2 of territory, but a very low density of population over most of the country, as the total population is only 33.5 million. Even this human/land ratio is illusory, however, since a third of the population,

Table 7.1 Selected characteristics of LDCs in three continents.

Country	Per capita GNP (1993)	Agric.(%)*	Urban (%)*	Pop. growth (%)
Latin America				
Argentina	7290	6	86.1	1.2
Chile	3070	15	84.6	1.6
Brazil	3020	11	75.2	1.9
Dominican Rep.	1080	18	n.a.	1.9
Ecuador	1170	12	56.3	2.4
Honduras	580	20	n.a.	3.0
Paraguay	1500	24	47.5	2.9
Peru	1490	20	69.8	2.1
Sub-Saharan Africa				
Burkina Faso	300	8	15.2	2.8
Cameroon	770	22	40.3	3.0
Chad	200	44	31.6	2.5
Côte d'Ivoire	630	37	40.4	3.8
Ethiopia	100	48	12.3	3.0
Ghana	430	47	34.0	3.2
Kenya	270	29	23.6	2.6
Mozambique	80	33	26.8	2.6
Zambia	370	29	42.0	3.1
South and Southeast Asia				
Bangladesh	220	33	16.4	2.2
Bhutan	170	41	5.3	2.2
India	290	31	25.5	2.1
Laos	290	51	18.6	2.9
Pakistan	430	25	32.0	3.1
Sri Lanka	600	25	21.4	1.3
Vietnam	170	29	19.9	2.4
NICs in southeast Asia				
Japan	31,450	2	77.2	0.4
Malaysia	3,160	16	43.0	2.4
South Korea	7,670	7	72.1	1.0
Singapore	19,310	0	100.0	1.9
Indonesia	730	19	28.8	1.8
Philippines	830	22	42.7	2.3

Source: World Bank Atlas 1993; UN Statistical Yearbook 38, 1990–91.
*Agricultural and urban populations as percentage of total population.

11.8 million, live in metropolitan Buenos Aires, and this one city houses 41 per cent of the national urban population. Population growth is slow, at 1.4 per cent annually, and getting slower. Urban population is increasing while the rural population is decreasing, in relative and in absolute terms.

It is worth prefacing the description of economic development with an account of the growth of the nation. At the time of independence from Spain in 1825 this was not a nation, but a collection of loosely linked settlements, indicated by the first post-independence name given to it, the Confederation of the River Plate. This had jurisdiction over a very limited area around Buenos Aires and up the river Parana. Beyond this, territory had to be won by conquering regional lords in the interior provinces, or by pushing back the Indian frontier. Both of these were accomplished in the nineteenth century, and the nation was formed. Throughout, the state intervened and promoted the developments that took place, incorporating new territories as a positive policy. A strong geopolitical stance was taken to defend this territory and assert its Argentine status (Morris 1996a). Final territorial advances were made in the 1880s, in Patagonia, and the 1890s, in the Chaco, to claim the two extremes, north and south, for the nation. Developmental policies since that time have increasingly sought to maintain territorial control as well as achieve economic growth in the regions. The central point of this history, from the stance of the present discussion of development, is that the state intervened from the beginning, that the state existed before the nation, and that only gradually did a sense of nationhood and civil society organizations to match come into existence. There is thus no question of whether the state can be held responsible for development over space. It was active from the outset, and this may be found to be the case in most of the Latin American countries, as in many of the LDCs (Morris 1996a).

At a national level, this country's economic development was spectacular up to the Peron period of the 1950s. From a backward colony of the South American empire of Spain, it became, in the course of the nineteenth century, a major exporter of raw materials to Europe, starting in the 1830s with wool, hides and salt meat, and adding to these frozen meat, wheat and other grains. On the basis of the new wealth, and with the cultural import of Spanish and Italian immigrants in the latter part of the century, Argentina also achieved a good level of general education and a reasonable provision of welfare services.

Spatially, the country's economy and society concentrated heavily around the capital city, where the main port was created and where the railway system had its hub. Buenos Aires was the main port both for exports and for the import trade into the country. Highly centralized government gave further impetus to this concentration at Buenos Aires. Economic development was eventually checked from the time of the Peron period, 1945–55, by an abrupt forcing of industrial development in a project of national autarchy like that of Spain. Manufacturing was developed, but at the cost of overburdening agriculture and checking its development. Argentina had already begun to develop substitutes for its industrial imports in the 1930s, when the northern hemisphere was unable to supply goods during the Depression, and during the Second World War, when lines of communication were cut off. After the war, this continued as import substitution industrialization (ISI) became a central policy in Argentina. Many industries were created on this basis, without a strong competitive advantage, but feeding on the large domestic market in and around the city. ISI policies meant a build-up of industries in and around Buenos Aires in the first stage up to and including the

119

1950s, since the early industries concentrated on final consumer products. Their best location was close to the largest single urban market in the country.

Another developmental dimension was the social. Argentina failed to develop an open society, and instead remained with a strong state apparatus and an elite which controlled it, and on the other hand, a growing mass of the poor, immigrant peasant stock from southern Europe. Frederick Jackson Turner (Hennessy 1978) had claimed that the US society was formed through the thorough and democratic mixing of homesteaders on the western frontier, as that country expanded. Argentina's frontier was more or less deserted, incorporated into great ranches rather than given out to homesteaders, and the few farmers there were temporary tenants, who mostly migrated to the great city of Buenos Aires before they could begin to create the structures of civil society in the countryside. A small landholding elite thus came to rule the country, and restrict economic change and access to resources for the many.

In recent decades, Argentina's developmental model has encountered serious problems. Following the earlier and easier stages of import substitution, from the 1950s the country started to move into the building of more advanced import substitution. In the terms of Dietz (1995), this meant moving from "horizontal" to "vertical" substitution strategies, producing not only consumer non-durables such as beer and clothing, but also consumer durables such as cars and television sets, cookers and washing machines. This led to two kinds of problem. One was the rapid exhaustion of the market, since economies of scale demand much larger markets than Argentina could muster for efficient manufacture of such items as cars and washing machines. In this way, the country started to provide its industries with heavy tariff protection from competitors, and they became inefficient. Another problem was the continued reliance on imported goods, since the inputs to such industries still came from abroad, and heavy imbalances in foreign trade were incurred. All Latin American countries became partially aware of these problems of external dependence, which would eventually produce a debt crisis that has slowed further development. Their solution, to form an internal grouping of countries in a Latin American Free Trade Area (LAFTA), was not a real solution, partly because it was never properly instituted, and partly because the enlarged market was still effectively a protected one, a Latin market. Current groupings, like the Southern Cone Common Market with Brazil, Uruguay and Bolivia, have yet to prove that they are different in content.

Within the development programme of Argentina itself, the first efforts at vertical ISI encouraged a yet further incursion into import substitution, and Argentina became a producer of a variety of goods, such as steel. Its major integrated iron and steel mill at San Nicolas on the River Plate above Buenos Aires started production in 1960. This was intended to rely on domestic raw materials, but there was never any domestic source of coking coal, and the iron ore mined from Sierra Grande in northern Patagonia was always high cost, the mine in fact being closed in 1990. This was not even the first steel mill; this had been at Zapla, near Jujuy in the far northwest, where a military government in 1943 had installed a mill based on local ores and charcoal instead of coke, from some 18,000 hectares of plantation eucalyptus. This mill was highly uneconomic, but it demonstrates the extremes to which countries concerned for self-sufficiency will go in defying markets. The steel industry is only one example; similar ventures in industries such as oil production, refining and petrochemicals, in timber, pulp and paper, aircraft manufacture, and in agro-industries such as cotton, sugar, and tobacco, were built up behind a wall of tariff protection. In the 1970s, investment was expanded into massive hydroelectric schemes, the most notable being the Yacireta scheme on the Middle Parana, which was not even built for 20 years because of political machinations, although large amounts of public money were spent. A number of hydroelectric projects were completed on the middle valley of the Rio Negro at the same time. All of these large projects were direct investments by the government, or by autonomous agencies which were in the public sector and removed from competitive pressures.

While industry was being sponsored by government, the farming regions of the interior suffered from population loss and a stagnation of the farming sector. Taxes on farm exports meant a loss of interest in the sector and a lack of investment. In addition, there was a large area in the northwest of near-subsistence farming and large estate agriculture, comparable to Spain's Andalucia. The ratio of income per capita in the northwest to that of Buenos Aires was 1:10, a far more extreme difference than in western European countries.

Forward development by Argentina, once checked in the first Peron era, was very limited in the period thereafter up to the advent of the Menem government in 1989. Hyperinflation and the inability to bring the public sector budget under control (partly because of the many huge projects in the interior) made overseas companies unwilling to invest in the country, and full opening to competition, externally through reduction of tariffs, and internally through privatization of the large range of state industries, was only achieved in the 1990s.

Regional policies

It was noted earlier that from the beginnings of Argentina's history there has been intervention by the state to control territory, and that expansion of the effective area of control was a major feature of the nineteenth century. Formal regional policy came much later, when from about 1960 there were a number of effective policies in operation as part of national planning, with the stated purpose of correcting the major imbalance between an overgrown centre and the rest of the country. In Friedmann's (1966) scheme, the Buenos Aires conurbation of around 10 million people could be regarded as the centre region combined with the upward transitional region, and the rest was largely downward transitional in nature, apart from the large southern zone of Patagonia which might be regarded as the resource frontier, as it had a small population and important resources such as hydroelectric power, sheep for wool production, and fishing potential.

There was no peculiarly Argentine form of regional policy. In the early 1970s, the problems of unbalanced growth were addressed partly through a growth poles policy in the Third National Plan. But as this was never followed through by later governments, only lasting for three years, it could not hope to have much effect. The poles were set up in the far northwest, at Salta, and in the northeast at Posadas, but several were in the south in Patagonia, which focused the growth poles on areas of national interest, for geopolitical reasons, but these were scarcely areas of urgent regional problems (Morris 1972). Argentina has always had worries over its effective domination of the national territory, and much effort has gone into defending the southern frontiers against imaginary enemies. Argentine claims to the whole of Patagonia are indeed somewhat dubious (Escude 1987), and firm control over the area was desirable to assert ownership. There have been other policies for industry, with a modest regional impact. Over the years, policies of industrial promotion (Morris 1992a) were the main force in redirecting economic efforts away from the capital city. From tentative beginnings in 1956, with the lifting of export taxes for southern Patagonia, the industrial promotion policy was extended and made more complex, with an array of tax remissions and other financial instruments used to encourage investment.

Most of the interior provinces became claimants on the industrial promotion programmes, with the strongest support for Tierra del Fuego, the Patagonian provinces, and Tucuman in the north. As in Europe, the effects of such programmes have been uncertain and never as strong as hoped for. It has been shown (Schvartzer 1987) that the main impact towards decentralization of industry away from Buenos Aires came, not from industrial promotion policy, but simply from the recession in Argentina during the late 1970s and early 1980s. With the opening of Argentina to trade, it was no longer necessary to have industrial bases within the country, and the lack of confidence in governments at the time meant a massive outsurge of investment from Buenos Aires to other countries. In effect, this produced a decentralization of industry, since the special aid given to some interior provinces meant that their industries were still worth maintaining.

Those industries that survived in the interior only did so through subsidies, and not based on a solid industrial infrastructure or efficient operations. In a country where tariffs were so high (over 100 per cent on many items of regular importation), lowering of the tariff barrier is a tremendous incentive, which may lead to location decisions that are wrong in the long term. It is worth mentioning a couple of the aided industries, as they demonstrate the artificiality of the whole operation. In Tierra del Fuego, because of its special aid status, which included freedom to import components as a free port, and a sudden need to bring in made-up items when the country moved from black and white to colour television, a national industry making colour televisions was established (Morris 1996b). This was no real industry but an assembly operation, using kits whose components were already built into subsystems of the whole, and only required the last assembly operation to be ready. No industrial skills have been built up in Tierra del Fuego, where the industry is still located, and the market is in distant Buenos Aires. Another example is in Chubut province, where a substantial textile industry has been built up at the small town of Trelew. Again, the advantage for the industry was the host of credits and tax remissions available from central government, together with special aid from the province itself. Again too, the industry is in reality poorly located in a remote part of the country, its vertical linkages being with Buenos Aires (materials come from the capital and return to it after weaving operations at Trelew), and there being few linkages between all the small factories in Trelew itself.

There have been other sectoral policies, including state ownership of industries, with a strong regional effect. Massive development of hydroelectric power has had some effect on the economies of distant provinces in the far north and south. National ownership of the petroleum and gas industries may have had a location effect, although private development of this industry is likely to have been more active. National ownership of transportation – railways and airlines – probably meant some reduction of the penalty for distant provinces, as market prices were not charged.

Summary

Overall, Argentina's development has been weak, and less than expected in the early decades of this century. Strong government intervention has apparently not helped, although there is a certain inevitability about state intervention given the early history of the country. A large set of policies was used in the 1960–90 period towards helping the interior regions, although without a strong effect, since the range of income differences between the highest and lowest provinces, per capita, is about the same as in 1970. There was, up to the advent of President Menem and new policies, a sense of failure about both national and spatial or regional development in the country.

There are three arguments about success and failure that may be summarized. One is the lack of openness of the Argentine economy, which has not helped it to succeed as a country that needed constant reference to export markets, because of its small domestic markets. Production for domestic markets or for the Latin market was always under protection.

Another argument concerns the role of the large territory and resources. On the one hand, in Argentina itself there is the claim that the country is overcentralized. Some aspects of the overcentralization may be measured by the extra-high costs of public services in Buenos Aires, or pollution and congestion costs. But these costs may well be balanced by the great economies of scale, whereby larger scale operations can be undertaken in the country's largest and richest market region. There are also external economies through each firm being set amongst neighbours that can supply services or components to it. Concentration of effort is beneficial, as we shall see in reference to the NICS, and as was the case in nineteenth century industrial districts. On the other hand, a reverse argument may be made. This is that the country's huge interior regions, and its resources, have encouraged the wrong kind of industries and set the country on paths it could not sustain, through encouraging development of the interior. This kind of argument may be valid for countries such as Venezuela, which have had oil wealth that they have not been able to manage successfully, which has caused a lack of competitiveness in other national industries. In the case of Argentina, no single resource, or group of resources, has had this effect, but the argument, following the discussion above, does hold some value. A variety of concerns for the interior have distorted national efforts away from the most efficient course. In the long term, this may not even be an equitable course, if national wealth does not allow for redistribution to poorer regions.

A third set of arguments concerns the political and social structure of the country. This country is a newly formed one which has had to struggle to achieve national stability, and whose centre has sought power and control over the territory. An aristocratic and often military-linked government has been ready to intervene constantly in the economy over the past 50 years. Intervention has been towards self-sufficiency, more than an ISI policy, seeking to ensure the country's capacity to provide for itself in all respects, including military defence through strategic supplies. There are territorial aspects of this policy, including the need to achieve regional development, to integrate regions into the centre, and to ensure they are not open to aggression from neighbours. Geopolitics here combines with economic considerations. As Street (1995) has amply documented, from the 1940s to the 1980s, continuous intervention by military or dictatorial governments served to limit academic freedom in universities which have produced a number of Nobel Prize winners, and has delayed technological innovation and economic development through its disruption of a stable learning society. There are "rents" obtained by this kind of intervention, in the maintenance of wealth among landed proprietors, of power in the nexus between these proprietors and the military, which have acted against the creation of an open economy and society. Such features contrast with the democratic origins of some other LDCs.

India

A country of subcontinental size, India is comparable to Argentina, but its demographic parameters are quite different, with 900 million inhabitants, a growth rate over 2 per cent annually (Argentina 1.2 per cent), infant mortality of 79 per thousand (Argentina 29), illiteracy at 52 per cent versus 5 per cent, and its rural population 74.5 per cent of the total, against Argentina's 14 per cent rural. The degree of concentration of population in one city is also much less. The population of Calcutta, is 13.3 million, but this only represents 5.8 per cent of the urban population.

This republic emerged in 1947 alongside Pakistan as the inheritor of British

India, a large country with an already massive population and difficulties in feeding this population. Compared with Argentina's 16 million people in 1947, India had 360 million in 1950, with only a little more land (3.1 million km² versus 2.8 million km² for Argentina). Much of this population was rural, so that while Argentina was already mostly urban, with a huge concentration of 5 million people in its metropolitan area, India was over 80 per cent rural. It has urbanized only very gradually since then. Population and its growth is a major check to any developmental process in this country, as in India there is no leeway of extra food resources or exportable materials to feed a suddenly growing population as the demographic transition starts. Modernizing pressures, in a country with limited resources and a dense rural population, can produce powerful negative effects in the environment. Although the matter will not be pursued in regard to this country, India is known (see Barke & O'Hare 1991) to represent well the environmental damage problems likely when more production from the land is required in an intensive farming system. Desertification affects substantial areas of the Deccan plateau, and deforestation and erosion/sedimentation are having startling effects in the forests of the Himalayan foothills.

Socially, the structure was perhaps less polarized than that of Argentina, but it also included an elite that emerged partly during the colonial times. Under British rule, the domestic elite had been intermediaries between the governing country and the governed Indians, acting in local government, English-speaking, and generally well-educated. In contrast to India in general, they had only limited links to the land.

There were two visions of India at independence. On the one hand was Gandhi's vision of a grass-roots level of organization, each village having some autonomy, each sharing its wealth in communal form amongst the members, through the agency of a *panchayat* or village council. Opposing this was Nehru's idea of modernizing development, neither the communism of China or Russia nor the open capitalism of the West. It involved heavy state intervention to promote and manage industrialization, since the private sectors, the great capitalists, could not be trusted to do this fairly.

In the event, and despite much propaganda and promises to respect Gandhi's vision, the government has led India predominantly down the path indicated by Nehru, with state control over a programme for heavy industry, to be carried through by a series of five-year Plans. In recent times these have been less important because of privatization, but in the early decades, the Plan was all-important. The process began with the First Plan in 1950–55, which attempted to give some balance between agriculture and industry. But by the Second Plan, only 3.6 per cent of Plan expenditure was on village and small-scale industry, while 17.5 per cent was on large-scale industry and minerals. Of the 3.6 per cent for small-scale works, most was spent not on Gandhian craft industries, but on small industrial estates for modern industry (Farmer 1993).

India has thus gone down the path of self-sufficiency in all possible products, to be attained by giving high protection to domestic producers, and by state ownership and investment in industry. Public ownership of the railways, airlines, post and telecommunications, major banks and insurance companies, atomic and all other energy sources, has been carried through since the early days, and only in the late 1990s are these national organizations being broken up. Joint ventures are used to give the state influence and control over steelmaking, cars, heavy machinery, hotel chains, and even retailing. Nehru's lead was followed in attempting to build up the heavy industries such as steel and heavy engineering, shipbuilding and chemicals. All of these industries were developed for the domestic market, and little attention was paid to the possibility of exporting. The growth achieved by this method was slight, always less than 3 per cent per annum, which was accepted stoically by Indian economists as "the Hindu rate of growth". India, prior to the changes of 1991, was stylized by Rohwer (1996) as "the tortoise in its shell".

There was a regional element to the policy, which was intended to induce maximum equality between the regions. This was added on to an already high level of control, so that India in the 1960s and 1970s presented a virtually unique case of the degree of state intervention in planning the economy (Bhagwati 1995). Thus, for example, in the textile industry, import substitution policy meant that the whole industry was protected from outside competition. But in addition, the large-scale sector was prevented from expanding too much, because there was a Gandhian interest in maintaining the small-scale craft and textile industry. Control was exerted through the system of licensing, whereby each firm had to have a licence for its annual amounts and types of production. This hamstrung the large-scale modern firms, and prevented them from expanding successfully into synthetic fibres. Beyond this control, there were further controls to ensure that the industry would be spread equitably amongst all the states involved in textile production, in a complete regional spread. This ensured a highly inefficient production system, as the more efficient producers could never expand to displace the weaker ones. Even more seriously, it encouraged a high level of corruption in the politicking by states and producers to gain licences for their production. Too much energy was wasted in "rent-seeking" activities, i.e. the search for influence with bureaucratic powers, and the search for power itself within an overgrown bureaucracy.

At the time of independence, India had a fair degree of inequality of income levels between the provinces, and this has remained at a similar level ever since, with few changes. Table 7.2 shows that the states with the great cities, like Maharashtra, with Bombay, and West Bengal, with Calcutta, have been the richest ones, and that the poorest have been states in the south and those of Assam. The differences are blurred by the large size of the states, which average poor rural and relatively rich cities, but they are probably less strong than those recorded for Argentina. Given its poverty, India's income differences are moderate and do not show massive increases.

In the Plans there is regular mention of the need to promote regional equality and the diminution of interregional differences in welfare and wealth. However,

Table 7.2 Per capita domestic product (in rupees) in Indian states.

State	1962–3	1982–3
Punjab	421	3418
Haryana	381	2873
Maharashtra	419	2634
Gujarat	413	2400
Himachal Pradesh	345	1967
West Bengal	420	1771
Andhra Pradesh	338	1713
Jammu and Kashmir	267	1705
Kerala	305	1689
Karnataka	327	1679
Tamil Nadu	365	1626
Rajasthan	289	1622
Assam	349	1596
Uttar Pradesh	258	1501
Manipur	172	1498
Madhya Pradesh	280	1423
Orissa	261	1339
Meghalaya	*	1308
Bihar	232	1120
Sikkim	n.a.	1079
Nagaland	n.a.	n.a.
Tripura	297	n.a.

Source: Brass (1991).

*Included in Assam; n.a. = not available.

the major programmes of the Plans have always been sectoral, with a regional impact that was uncertain. The licensing system mentioned above in connection with textiles merely meant that no regions could claim special attention or have industries focused upon them. Special attention was given to providing industries in some peripheral states, and Bangalore was a beneficiary, with an integrated steel industry and help for its booming electronics industries. Because of the wide range of new industries set up in India, and the licensing system, there was scope for spreading industry to many new sites, and most states have enjoyed industrial growth.

Instead of the three great cities of Calcutta, Bombay and Madras, dominant in colonial times, there are now some 50 large cities with substantial industrial development. Rather than textiles and food industries, which accounted for 70 per cent of manufacturing when Nehru set up the programme, there is a host of large industries in cities all over the country (Fig. 7.1). The primary concentrations are well distributed, in the regions of Calcutta (jute, engineering), Bombay–Poona (cotton textiles, engineering, chemicals, electronics), Ahmadabad–Vadodara

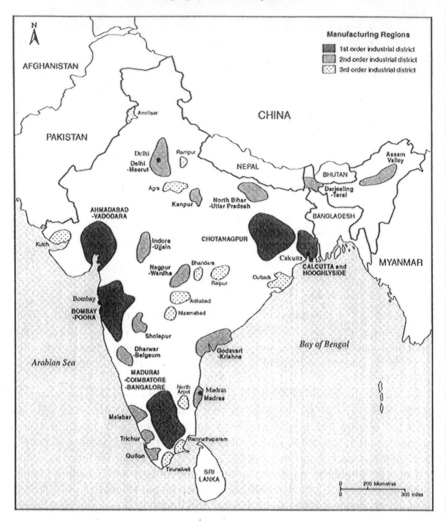

Figure 7.1 Indian industrial districts. The three different orders are categorized by Indian geographers on the basis of number of workers. (Based on Dutt & Geib 1987.)

(cotton textiles, petrochemicals, iron and steel), Madurai–Coimbatore–Bangalore (cotton textiles, electronics, electrics, aircraft, machine tools) and Chotanagpur (the traditional heavy industry region, with iron and steel, engineering, aluminium manufacturing and vehicles).

The central state played an important role in spreading industrial growth to all parts; chemical industries were widely dispersed, with a slogan that called for a fertilizer factory in every state of the union. Thus, although development is not even, India could be regarded as having a geographically well-distributed industrial development process, which may be said to be advantageous, other things being equal. The heavy environmental and congestion costs of over-centralization are reduced by the location of textile mills, for example, throughout the country (although still with major concentrations in the cities of Bombay, Ahmadabad, Madras and Calcutta). As the individual states have taken over much of the industrialization process, advertising their own industrial estates and competing for incoming firms, the spread of development is increased by the political decentralization. In terms of achievement to date, however, India has a dismal record.

One point might be made regarding the spatial effects of this striving for equality, including spatial equality. Given the kind of industries supported by the Indian state, such as steelmaking, it seems unlikely that a dispersed pattern of investment would be advisable. Other industries too, such as those based on human skills, are from evidence elsewhere highly concentrated, and dispersal merely poses a barrier.

Agriculture

Throughout the industrialization process, agriculture could not be forgotten. Apart from Gandhi's ideas for an idyllic Ruritania, it was necessary to support the population, still largely rural, and still farming at subsistence level. Agriculture has certainly had this sustaining function for India's 850 million people in the early 1990s.

A most notable success has been that of wheat production, up from 6.6 million tons in 1949–50, to 50 million tons per year in the 1990s. Other grains were less striking in their results, and the success with wheat was partly due to its role in "green revolution" technology, which had less impact on some traditional crops of marginal areas, such as barley, millet, sorghum and the bean family.

However, helping to keep a rising population alive is not enough to promote development. In the LDCs, as in the early development of Japan in this century, agricultural surpluses were useful contributors to savings by the nation and by farmers, later invested in services and industries. In India this has not been possible, despite the introduction of "green revolution" technology. "Green revolution" technology, the use of high-yielding varieties (HYVs) of certain crops, and their combination with fertilizers and ideal watering to produce high yields, has been criticized mostly in terms of its divisiveness between rich and poor farmers, the latter being unable to afford the expensive inputs such as fertilizers and herbicides needed to grow such crops. But we might note Lipton's study (1989) showing a high level of adoption amongst poor farmers. The use of HYVs was certainly divisive through being usable only in certain areas. The best wheat area is that of the northwest, in the states of Punjab, Haryana and western Uttar Pradesh. HYVs were found for rice, which extended the area of application, but for the flooded areas of the deltas and the dry areas of the interior, there were no easy adoptions.

Perhaps more importantly than any critique of the "green revolution" or other farming and rural programmes, is the fact that farming and rural development has had little effect in stimulating the development of the economy generally. Throughout the Independence period, production has been more or less enough to meet demands. At no time was there a crisis on a scale big enough to redeploy resources and concentrate on any specific strategy. No major cash crops have been able to command strong support to the stage of starting exports and bringing in foreign exchange, which might have been put into local wealth and thus into regional development.

Nor were there crises sufficient to stimulate a thorough-going agrarian reform. Reforms have taken place, but their effect has been softened by modifications and exceptions so as to make little overall difference to production. Most farmers have also been kept at a poverty level so low as to give no space for training, preparation of the younger generation, or improvement of skills in non-traditional areas. Surplus population has migrated out to the towns, but at a modest pace only sufficient to take away the surplus and still leave a high ratio of population to farmed land.

The case may thus be made that either a negative crisis (famine or severe shortages) or a positive one (surplus production for sale) would have promoted change in the rural economy; but that in the event, production just increased in line with the increase of population. Agriculture and the rural area for India did not thus represent a major resource that could be used to finance development, as happened in Argentina. Agriculture fed the country and little more. On the other hand, India does not have the large open spaces that are so tempting for investment in the South American countries, and so wasteful in most of the regional development plans. Instead, the Indian regions are well populated and represent potential markets. A spatially distributed development process would be useful by bringing more of these regions into the national market, and making this market bigger, allowing economies of scale for the great industries. This does not mean spreading out all kinds of industries to all regions, but moving towards a more dynamic economy in each region.

Interpretations

Comparing with Argentina, the history of economic development efforts in India is similar, with emphasis on mass industrialization, which was to be undertaken by the state to ensure its fulfilment. But whereas Argentina is characterized by heavy centralization in one city, India has many large cities, and a very large rural population. India, however, is not a success story, and its level of income GNP per capita at around $300 in the early 1990s, compared with Argentina's $7000, is an order of magnitude lower. Nor has there been any period of rapid growth of this income level to show change. Evidently, it is possible to be poor over a well-distributed system, to share the poverty, as much as it is possible to concentrate wealth in Argentina. One major difference between the two countries is the rapid population growth in India, which makes it much harder to attain per capita growth of income. A 2 per cent population growth rate means that production change has to include this 2 per cent before achieving any positive per capita changes.

The problem to be explained with India is its very poor performance in terms of the growth of per capita income, still lying at around $300 in the 1990s. This is one of the poor countries, not a candidate for election to the middle-income group. A first answer is its concentration on heavy industries, which would take a long time to finance and bring into production (Auty 1994). But this is not a full answer, for India has concentrated on heavy industry since Nehru's initiative in the late 1950s under the Second Five-year Plan. By now it should be reaping full benefits from these industries. A related point is India's following of the ISI route of industrialization, protecting many infant industries, which can be inefficient. This has much truth, but in such a large country, the markets are big enough to grow efficient industries of any sector, and to introduce competition over time. Another kind of answer comes from the relationship of economic development to demographic change. India's high rural population means that traditional families of six or more children are common. Population growth is still over 2 per cent annually, absorbing a lot of the growth of output. Argentina has not had this problem. Population growth has also clouded the advances that were achieved with "green revolution" innovations in farming.

A final explanation is that the Five-year Plan type of development has been subject to constant government intervention, with many public companies (in 1961, 6000; in 1993, about 18,000). The public sector employs over 70 per cent of all workers (Stern 1993: 211). Of itself this is not a damning situation, although it may lead to inefficiencies from lack of market disciplines. But in India, control of the public sector is by an urban middle-class elite of bureaucrats with English-language training, a self-perpetuating group not subject to controls themselves. This means that corruption and what may be called political "rents" (i.e. profits due to the special powers of the group) are prevalent and limit any economic success. The problem is thus not just one of inward-looking development, but also the chance this gives for mismanagement by particular social groups. A social factor must be added into the economic mix.

All this criticism refers to past performance; from 1991, India has begun to adopt more open policies, and to eliminate the damaging licensing system, as well as to attract foreign investment rather than prohibit or restrain it (Rohwer 1996). Thus there is new investment by Western countries, and by East Asian neighbours such as Singapore, into the region of Bangalore where the electronics industry is in rapid expansion. India's huge domestic market provides an instant base for changing its dismal past record into future success.

The Soviet Union and Russia

Although commonly excluded from discussion of developmental matters, or treated as a separate entity corresponding to different ground-rules and control systems, the Soviet lands do provide useful support to the theses advanced in respect of poorer countries generally, the concentration of development, and the problems imposed by large territories. The ex-USSR also provides evidence on another score, the relative success of a command economy, versus the market economies or regulated market economies of the West. This case does not represent a parallel to the east Asian economies, nor to the other, less successful LDCs. It is a case *sui generis*, but one that has features in common with other world regions. It has also been focused on in the past because of the specific and repeated claims of the Soviet leaders that communism would eliminate inequalities between different groups and regions within the Union. These claims, and some features of its development, are shared with mainland China in the period from the 1949 communist takeover to the present.

By 1990, on the eve of disintegration, the USSR was composed of 15 Union republics, dominated by Russia with a population of 140 million, 57 per cent of the total, and 75 per cent of the land mass. This land area, of 22.3 million km², was the largest effective political unit in the world, larger than the whole of North America including Canada, the USA and Mexico.

Rather than a single country, the USSR might be classified as an empire: first a Russian empire constructed from the late fifteenth century onwards, with the absorption of first the Volga tribal groups, then the Ukraine (seventeenth century), the Caucasus states in the early nineteenth century, and Central Asia mostly in the period from 1862–80. In Siberia there were no organized states to defeat, and after exploration reached the Pacific by 1689, colonization was gradual through the nineteenth century.

This vast land was an empire in cultural terms because of its ethnic diversity, but a single country in terms of forming a single territorial unit, compared with the west European overseas empires reaching to distant overseas colonies. Within the USSR, the geography may be simply divided into three regions. In the west, there is an abundance of human skills and there are the most technologically based industries. East of the Urals, there are abundant resources, including the forests of Siberia (three-quarters of the Soviet total), oil and gas, mostly now coming from Siberia, coal in the southern Siberia basins such as Kuzbass, copper in the Urals and Kazakhstan, tin in northeast Siberia, and precious metals, gold and diamonds. South of Siberia lies central Asia, the region with greatest population surplus since it has rapid population growth. Central Asia is not rich in minerals, but has an important agricultural production.

The regions

Regional policy has been with the direct aim of colonizing and economically opening the east, a move comparable to Brazil's drive to the west. Rather than a means to an end, this has been the end in itself. This kind of aim is little different to that of Argentina or Brazil, seeking to consolidate central control over peripheral regions, and represents a geopolitical strategy for the effective growth of the nation state in cases where the state is a somewhat artificial creation that antedates that of the nation. In addition, an aim which came in with communism was the elimination of inequalities between different groups, sometimes regarded as the "nationalities question", because the regional inequalities are in large measure inequalities between different ethnic groups, specifically between Russian slavs and other peoples. After 1972 this inequality was stated to be no longer a problem (Liebowitz 1991), but evidence is that it was never resolved and has again assumed greater prominence (Kaiser 1991, 1997).

There was no regional planning in overt form, however, for most of the Soviet period. Only during the Sovnarkhoz period of 1956–64, under Khrushchev, were there efforts to use a directly regional structure to make investments. For the most part, developmental efforts were conducted through the sectoral ministries in Moscow, which tended to centre on big projects, such as the creation of an iron and steel industry in the Kuzbass region of southern Siberia, which would bring along with it commitments to open mines, build railways, or construct workers' housing. This approach would have an uncertain effect on inequalities. An immediate comparison may be made between this and the developmental tendencies of Latin American countries such as Brazil, involved in large projects for opening the interior.

Some effect on disparities has been through policies of providing equal or higher wages in the east, to compensate for more difficult climate and environment generally. The higher wages in Siberia do not wholly balance the poorer production figures, but do help in this direction. There is also evidence that higher levels of welfare are provided in some eastern regions, so that on social measures such as schooling or availability of doctors, there are only small differences between the regions.

Interregional disparities in the Soviet Union are large. There are major difficulties in the interpretation of official data on the subject, as GNP measures and their like are not used. One approximation is the level of industrial output per capita. In 1926–7 (Dmitrieva 1996) there was a ratio of 5.22 to 1 between the highest unit, Russia, and the lowest, the Kirghiz republic. The range in 1988 was 4.33 to 1 between Estonia and Tadjikistan. Estonia was not a member of the Union in 1926, and had it been so the earlier value would have been still higher. There is thus some diminution of the differences over time, but even the 1988 figures indicate a very great gap. Some of the gap must be expected over such a huge territory; the larger the unit, the more likely are differences to be found. But the values seem to refute at once the idea of equality. Most other output indices for the recent past show comparable levels of difference, about 5 to 1. These are based generally on primary and secondary sector production, since the service sector, producing non-material goods, was not regarded as contributing to wealth in Marxist and socialist economics. If services had been added in, it is likely that the differences between regions would be larger still, because large urban regions such as Moscow and Leningrad and their surrounds have a high development of services compared with some poor regions. A further increase of the overall differences between republics in the USSR would be obtained from using world prices for goods rather than those of the Soviet Union itself, where all prices were artificial, being set by the state. This kind of policy meant that oil and gas and most minerals were underpriced, and thus the output for Russia itself was artificially low. Using market prices would mean increasing Russia's output differences over the other republics.

In recent decades the differences have ceased to decline, and since 1960 have shown an increase related to the decline in developmental impulses towards Siberia and the termination of large projects in central Asia (Dmitrieva 1996). Most rapid growth and development has been, not in Siberia nor in the heartland of industrial Russia, but in the regions marginal to this latter, along the Middle and Lower Volga and in the Urals.

A case study: Central Asia

Russia may be said to have two peripheries. First, to the east, there is a resource frontier with a thinly scattered population. A second one is the ethnic periphery, whose main components are the Caucasus and central Asia. Examining central Asia, most commentators have argued that this has been a kind of internal colony of the USSR. The countries of this region, including the Turkmen, Uzbek, Tadjik and Kirghiz Republics, certainly have a lower standard of living than in the centre, ranking in a group fifth out of six classified by Dmitrieva (1996) on the basis of household income and retail volume data, as well as social welfare criteria such as schooling and medical care. The case generally made (Liebowitz 1992) is that Russia purposely developed the Central Asian region as a provider of primary goods, principally cotton, to the detriment of other production, and in a dependent fashion. The impetus for this development came from Moscow, including investment capital, skilled personnel, and markets for the cotton. As in dependency relations elsewhere in the world, pre-existing industries, here artisan work from the long inheritance of a region on the trade route between the Orient and Europe, were neglected and left to decline.

Central to the development was the expansion of irrigation of the deserts around the Syr Darya and Amu Darya rivers, which had covered an area of 2 million hectares in 1913, but had grown to irrigate 7 million hectares by 1990 (Sinnott 1992). This expansion, in its later stages from the 1960s, was to have involved huge hydrological schemes, notably the diversion of northwards-flowing

Figure 7.2 Geography of the Central Asia region of the former USSR.

rivers, in Europe to the Volga, and in Asia from the Ob to the Aral Sea area. As these schemes have not materialized, the take-off of irrigation has caused major ecological problems in the region and particularly in the Aral Sea.

Because of reduced flow, the area of the Aral Sea has halved between 1960 and the mid-1990s, and the volume been reduced by two-thirds (Clem 1997). The shoreline has retreated over 125 km, and the Sea divided into two halves from 1989. Most of the fish have died from the heightened salinity. Around the shores, the cold dry winds of the steppe blow off salt and dust from around 30,000 km² of exposed lake bottom, creating new health problems for the population living near the Aral. Inflow from the Syr Darya river has ceased altogether from 1978, and the Amu Darya carries very little water; what there is has been polluted by pesticides and is low in soil nutrients. Around the delta areas of the rivers, there has been an ecological change from a rich forest mixed with fresh-water ponds, to a poor salt-resistant vegetation and wildlife. It can be argued that the ecological disaster portrayed here is a further aspect of dependency; in other words, it is because of the dependency on the centre that wrong decisions were taken, neglecting ecological problems.

There are some comments that temper the dependency arguments made so far, however. An important minority group in Central Asia seems to have benefited from the cotton monoculture, and to have supported it. They were the owners of land used for fruit and vegetables, and cattle farmers, often illegal, outside the main cotton farms, who were able to produce at high prices for the local markets (Dmitrieva 1996). For this group, and for others with extraordinarily productive smallholdings, outside the *kolkhoz* system in central Asia, rural incomes were quite high and the cotton monoculture did not represent the repression of an impoverished peasantry. A further point is that the investments coming from Russia did provide some improvement in the physical infrastructure and welfare provision in the region, and without them, central Asia's economy might have been more comparable to its poorer neighbours in the Muslim drylands reaching from the Middle East to Pakistan.

Recent developments

In the years from 1985, there has been a progressive liberalization of the ex-USSR. First, under Gorbachev, there was a gradualist approach, opening up individual state enterprises to market rules and allowing some competition. This work, at the microeconomic or firm level, was succeeded by a more rapid and complete change from December 1991 when the USSR itself was dismantled, being followed by the Community of Independent States (CIS), and leaving the Baltic states outside the grouping.

Under Yeltsin from 1992, macroeconomic changes began, including the dismantling of foreign exchange controls, and the institution of competitive markets rather than a command economy. This period has also seen the wholesale privatization of state enterprises, including some 14,000 enterprises moved to private hands over 14 months, and decline of state spending from two-thirds of total GNP to one-third by 1995. The cooperative and state farming sector has also largely been privatized.

All these changes have left Russia as a relatively poor country, especially because of the disruption of industries through changes in their organization and economic environment. Russia, Belorussia and the Ukraine had GNPs per capita in 1994 of between $1500 and $3000, which places them not in the poorest category of countries, but roughly on a par with the countries of South America. This level sits uneasily with the earlier pretensions of world-power status on the part of the Soviet Union.

On the other hand, some of the changes may be seen as the inevitable result of overreaching plans and programmes by the central government, which it was thereafter forced to retract. Russia's present poverty may be a record of false leads from the past. Investments in Siberia and Central Asia are likely to be greatly reduced in value, once the CIS is opened to the West and to competition, which will force the costs of eastern development to be taken into account. These regions are likely to suffer in the near future as their support is diminished. Some areas of European Russia may also suffer where they have been built up around a single industry such as steel, or where large-scale industry and mass production methods have been adopted for consumer goods, out of consonance with Western demands for highly differentiated and fashion goods.

Russia's current poverty and the need to invest in restructuring also means that there can be no early solution to some of the major environmental problems, which have never been properly addressed (Pryde 1997). Besides the Aral Sea question, discussed above, there are major regional pollution problems. Amongst the most notorious are those associated with Chernobyl and nuclear fallout, with the heavy industries and air and water pollution in the Ural Mountains, and with Lake Baikal in Siberia where forest-cutting and pulp mills are still active at the lakeside. But many of the cities suffer from air and water pollution, and the Volga River has pollution problems along most of its length.

CHAPTER 8

The newly industrialized countries: South Korea, Taiwan and Indonesia

Amongst the countries outside the older advanced countries of western Europe and North America, most attention has focused on the NICS, the newly industrialized countries, as the success stories in development. There are important differences in the way these countries have developed compared with the overpopulated ones of Africa and Asia, and from the thinly peopled ones of South America.

One macro-region of the modern world undergoing rapid development is that of southeast Asia. Here, as Table 8.1 indicates, a set of countries has achieved a similar kind of growth and development over time, with powerful growth of GDP from the mid-1960s, and an improvement in living conditions indicated by the high life expectancy and low population growth rates. The analysis made here is mostly in terms of two countries, but it should first be noted that a whole system of industrial development has been occurring in the coastal and island territories of east Asia. As noted by Park (1992), there was a sequence of development impulses, spreading from one country to another through the region. These impulses started with Japan in the 1950s; they spread to the NICS of Korea, Singapore, Hong Kong and Taiwan in the 1960s; to Thailand, Malaysia and Indonesia in the 1980s; and finally, in the 1990s, the spread has reached the mainland of China, to the coastal provinces from Guang Dong (Canton) in the south to Shandung (Shantung) in the north.

The similarities in these growth impulses are, first, that they have all been based on industrial growth, not on primary sector products. Secondly, and linked to this non-reliance on primary exports, and because of the needs of industry, there has been a great reliance on human resources, education and training for the population. Thirdly, they have all employed an outward-oriented strategy of growth. This does not mean simply export-oriented, but also the use of foreign capital, foreign technology and innovations, through a rapid learning process. Outwards orientation has also involved the building-up of links between the various countries of the region.

It must be noted that the influence of Japan is important throughout the region, in moving out many of its labour-intensive industrial operations to neighbouring countries with lower costs, and in investing financial capital and making direct foreign investment in the economies of several of them, notably Korea. Japan has

Table 8.1 GDP and population growth figures for NICs compared with other countries.

Country	GDP growth		Population growth (% p.a.)	Life expectancy (years)
	1965–80	1980–90		
NICs				
Korea	9.9	9.7	0.5	79
Taiwan	10.2	7.8	1.4	75
Hong Kong	8.6	7.1	1.8	78
Singapore	10.0	6.4	2.1	75
ASEAN group*				
Thailand	7.3	7.6	1.8	66
Malaysia	7.4	5.2	2.6	70
Indonesia	7.0	5.5	1.8	62
Philippines	5.7	0.9	2.4	65
Developed countries				
USA	2.7	3.4	0.9	76
New Zealand	2.4	1.9	0.8	76

Source: World Bank Atlas, various years.
*Association of South East Asian Nations.

also imposed something of a style of development, in the elements listed above, which has been proved successful. As central evidence for their rapid economic progress, the growth rate of GDP for some of the countries in critical periods may be noted.

The best-known cases are the East Asian "Tigers", the states of South Korea, Hong Kong, Singapore and Taiwan. Hong Kong and Singapore are city states with tiny territories; as this book is concerned with spatial structures, we focus on the larger states of Taiwan and South Korea. But it might first be noted that Taiwan and Korea are also small on most scales; South Korea is 98,500 km² in area, and 60 per cent is upland with very low potential for use. Its population is large for this land area: 43 million in 1991. Taiwan is also a small country with forested mountains and few resources; it has only 36,000 km², less than a tenth the size of California, but with a comparable population (20.4 million in 1990). Thus all the east Asian Tigers have large concentrations of population on small territories.

South Korea

Geographically small (Fig. 8.1), this country is still smaller in effective resources, with 60 per cent of the land being mountains with a severe winter climate, only suitable for forestry. Population is heavily concentrated in the large urban centres, notably in Seoul which houses over 40 per cent of the national population, and in the southeastern region around Pusan. In the past and up to relatively recently, there were dictatorial governments and a lack of any real democracy. Earlier this century the country of Korea, including the North, was a colony of Japan between 1910 and 1945. In that period, Japan did not actively develop the potential of the country, which was made to produce raw materials in exchange for Japanese manufactured items.

After the Japanese period there was a stage up to 1960 when Korea tried the ISI type of policies attempted in Latin America and endorsed by current thinking for the development of Third World countries. These policies did not bring rapid development, although this was in any case a time of recovery from the Korean War, when a million lives were lost along with two-thirds of the national

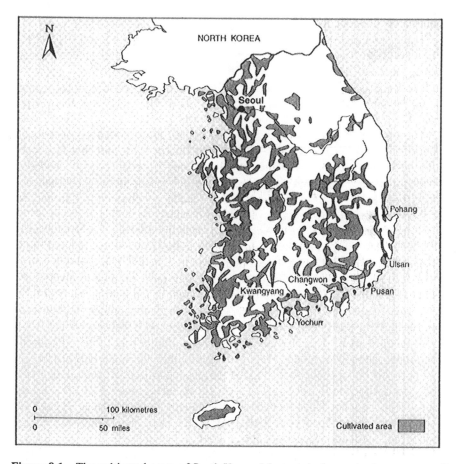

Figure 8.1 The cultivated areas of South Korea. Most are in the west, plus a scatter of valleys elsewhere.

productive capacity. From 1960, however, there was a radical change of policy, and rapid development led by manufacturing industry. In this second development stage, 1960–73, it was light industry, textiles, clothing and plywood that were favoured: medium-level technology industries that could readily be put in place. The market was initially the domestic one, so that there was indeed an import substitution function, but success soon made this an export function for many products. In a third stage, from 1973–80, there was a shift to heavy and chemicals industry, including petroleum and petrochemicals, rubber, base metals, heavy engineering and shipbuilding (Kwack, Ch. 3 in Lau 1990); in these industries, Korea was successful in penetrating world markets. The heavy industries were immediately useful to the economy of South Korea, through the forward linkages to light industry already in place. A later stage over the period 1980–87 was of economic liberalization, allowing more freedom for trade in both directions, and moving towards high-technology industries, consumer electronics and semi-conductors manufacture.

Trade, it should be emphasized, is still not totally liberated, and in particular, there are trade bans on many products from Japan; rice production is also still protected. Over the past few years, the changes have been less economic than political and social, with a move towards democratization (Herd & Jones 1994). It might be said that South Korea is now traversing the social and demographic transitions, following its rapid economic transition.

The choices of industrial foci were not casual, but planned by government. While the government did not own industries for the most part, it was a highly interventionist government which stimulated and guided industry constantly, through five-year plans and annual budgets which specified the industries to be targeted. Over the whole of the 1960s and 1970s, the government was military and was able to exert strong control over private industry.

Competitiveness in export markets was maintained by the "openness" of the economy which forced efficiency in all industries. However, as pointed out by Toye (1987: 87), there was no truly open economy, especially in the early years of development, and South Korea maintained strong tariff protection for its infant industries, with quotas and physical prohibitions on some imports. The tariffs have since descended, but they were used for a critical short period. What was always true was that exports would have to compete on world markets.

In South Korea the rural sector did not have a large role to play. Perhaps of some importance was the occurrence of agrarian reform in the period 1948–52, which redistributed about 80 per cent of the tenant-farmed land and eliminated all absentee landlord farming. This left virtually all farms under 3 hectares in area, however, too small to generate capital development. It would appear that the very absence of a large, important agricultural sector in modern South Korea is an advantage, not disturbing plans in other areas.

Throughout South Korea's development there has been a strong dependency on Japan; Japanese direct investment was made in factories in South Korea, as well as some financial investment. Korean industries grew with Japanese technology, and imported Japanese components for assembly and return to Japan. In recent years, Korean industry has invested in similar fashion in mainland China.

In all of this development surge, regional issues were kept at a low level of priority. Had primary products and agriculture been seen as vital, there might have been an effective regional policy, but the emphasis was on urban industrial growth centred on two core regions, around Pusan in the southeast, and Seoul in the northwest (Fig. 8.1). A huge concentration of population in Seoul makes this one of the world's largest cities, estimated at 19 million people in 1995. There were decentralization policies to move industries out from the two older centres, including tax incentives for relocation, the provision of growth poles, and some help for agricultural prices (Auty 1990). The growth poles, located in the southeast to help the weaker of the two main centres, Pusan, have been successful in production terms, achieving rapid growth for the region where they were established in the 1970s. The growth, however, still did not outweigh the very rapid growth of Seoul itself.

The larger growth poles were based on heavy industry complexes (other smaller ones were really only industrial estates), and petrochemicals, steel and machinery poles were set up. Obviously, as in the West, these heavy industries only achieve limited direct employment, but they benefited from being located within 100 miles of the big centre, Pusan. Major centres were at Yochun (petrochemicals) and Changwon (diversified machinery) in 1974, Pohang (steel) and Ulsan (petrochemicals, Hyundai machinery) in 1975, and Kwangyang (steel) in 1982. It was important for the success of these big centres that they had good access to other industrial centres, forming a network centring on Pusan. These poles are linked partially to the kind of firms that grew in South Korea. The four largest of these, Hyundai, Samsung, Daewoo and Lucky, each had a turnover size larger than the combined top ten in Taiwan, which relied on small firms. Strong government intervention supported these large conglomerates with many products.

During the period from the 1960s to the 1980s there were major regional changes, involving a reduction of interregional economic differences of considerable magnitude. In 1968 there was a 3:1 range of gross regional product (GRP) between the richest and poorest regions. By 1983 this had been reduced to 1.7:1. Such changes were not really attributable to the growth poles *per se*, so much as to a massive population shift from the rural regions to the cities, and especially to the growth zone around Seoul. The northwest core region, with 8.89 million people in 1970, had 28.3 per cent of the national population; by 1985, it had 15.83 million, or 39.1 per cent of the national total. The southeast region with its growth poles had seen some population growth, from 9.56 million to 12.08 million, but its proportion of national population fell to 29.8 per cent. Urbanization of the population went on at a rapid pace, with the urban share increasing from 41 per cent to 65 per cent of the total population over 1970–85. This had a great effect on regional product disparities, because the urban population was much better paid. There was a support system for farm prices, particularly for rice, but a central structural problem of small farms (average size 1 ha., and legal maximum 3 ha.) meant that farm income could never be raised much for the poor. A basic message must be that South Korean success owes much to concentration on urban industrial development, and to the fact that its rural sector has not posed too much of a drain on the wealth produced by the main strategy.

Taiwan

A second example of an NIC is Taiwan, called the Ilha Formosa or beautiful island by Portuguese explorers. This thinly populated backwater of China was suddenly projected into importance in 1949 when the Communists took over the mainland and forced a massive migration of their opponents, the Nationalists or Kuomintang, to the island. It too followed a Japanese model of rapid development with a state presence behind private firms, in a context of poor natural resources. But it also illustrates some important differences from the other east Asian Tiger economies.

Some of the elements in its geography and history are comparable to those of South Korea. Like all the NICs of the first phase following Japan, it has a relatively small area of concentrated settlement and activity. About two-thirds of the island is rugged mountains, in four sub-parallel ranges, lying mostly in the east of the island (Fig. 8.2). These mountains are of easily eroded shales and sandstones, uncultivable for the most part because of their slopes, and necessarily kept forest-covered. Forest covers 65 per cent of the island, mostly broadleaved evergreen trees of mixed species, overused in wartime and now kept more for conservation. There are few minerals worth exploiting, and no oil. In the west there is a low coastal plain with fertile soil, and most of the agriculture and population (66 per cent), although with only 22 per cent of the area.

Like Korea, Taiwan was a Japanese colony, in this case from 1895 to 1945. Before that, from 1683, it had effectively been an internal colony of China, supplying rice to Fukien province when that province experienced shortages. Under the Chinese there had been agricultural exploitation, and the growth of social division between landowners and labourers. Japan, in contrast, did make some investments, including those in infrastructure, roads, schools and hospitals, but did not encourage strong industrial development as Japan itself was the supplier of industrial goods to Taiwan (Hsieh 1964, Myers 1990). The beginnings of entrepreneurship were found in the many small industries processing sugar-cane, hulling rice, and the like. There was a first modest essay into hydroelectric power production, with the building of the Sun Moon Lake project in the mountains during the 1930s, to support aluminium, chemicals, and steel alloy industries. Taiwan's role for Japan was, however, to produce sugar and rice, and 95 per cent of its sugar and 52 per cent of its rice went to Japan (Ranis 1995). Rice production became highly efficient under the Japanese, who doubled the output

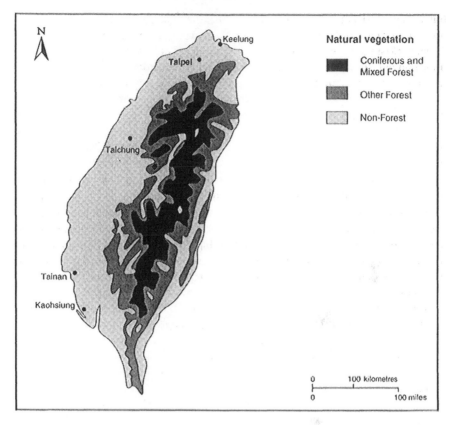

Figure 8.2 The natural vegetation of Taiwan. The vegetation patterns correspond roughly to topography and land use. The forest lands are largely mountainous and unsuitable for farming. The cities are on the western plains.

using irrigation, fertilizers, and improved weed and pest controls. Dependency did not mean no social progress, however; 60 per cent of the population was literate by the early 1950s. On the other hand, landownership in the island remained highly skewed, with a few landowners dominating the whole structure and many landless labourers.

After the exit of the Japanese, millions of migrants came to the island from mainland China, including many merchants and entrepreneurs displaced by the Communist takeover. The Kuomintang government saw the need for tight control over the economy, aware that part of their mainland defeat had been due to hyperinflation and lack of control in the 1940s. Thus an ISI policy was followed during the 1950s, up to 1958–9, when the change towards a more outward-oriented policy began. During the 1960s and 1970s, Taiwan achieved massive growth rates in production, as Table 8.1 demonstrates, shifting it out of the poor countries group altogether. In agriculture, Taiwan benefited from agrarian reform between 1949 and 1953, underwritten by the USA, building up a relatively prosperous small-farm economy. First the excessive rents were reduced to a reasonable level, then security of tenure was given to tenants. Finally, landlords with tenants were generally replaced with owner–operator farmers, and co-operatives were formed to help these farmers market their goods (Reitsma & Kleinpenning 1985: 360–87). Development after the Japanese departure thus benefited from a good physical and human infrastructure built up during the early part of the century, and an energy which was released once social inequalities were overcome. Most writers have emphasized the beneficial role of an agriculture that was productive and able to help in savings during the transition to an industrial society. What was more important was probably the social transformation achieved through elimination of the large-scale landlords.

As elsewhere, the development process was multifaceted, with an increase in savings rates from under 10 per cent in 1955 to 30 per cent in 1980. There was an equalization of incomes between social groups, from an excessive ratio between the top and bottom 10 per cent strata of income of 20:1 in 1953, to 4:1 in 1980. Without regional data, we cannot be sure that this also implies an interregional convergence, but this is a likely situation. Demographic transition was also achieved, with a 4.0 per cent birth-rate up to 1950, typical of LDCS, reducing to under 2.0 per cent by 1984, and a population growth rate of 1.5 per cent in this latter year.

Government intervention

As in South Korea, industrial development was a key feature in the overall economic development of Taiwan, and this was very largely by the private sector. In China there had been a strong tradition of state interference in the economy, and this was taken to Taiwan, where 48 per cent of basic industry was publicly owned in 1960 (Myers 1990: 43), but by 1981 this had become about 80 per cent private. During the 1980s the public sector has been reduced still further (Ranis 1995). Government intervention was always significant, but it is important to understand that this was not in the face of, or against, the private sector, but in support of it, including foreign firms in some selected sectors that were allowed to remit profits to home countries. Economic planning was formalized through national plans of five to ten years in length, specifying which industries were to be favoured (Selya 1993). In addition, the government invested in major projects such as roads and railways, ports and power plants, concentrating its resources heavily in different time periods. Evidently this kind of policy, like that of South Korea, does not correspond to a liberal view on development. On the other hand, it does not follow the dependency logic, of deciding that outside influence is bad and therefore trying to isolate the country from such influence.

The types of industry were somewhat different from South Korea, with much light industry. Few large firms were brought into being, contrasting with the South

Korean conglomerates such as Hyundai and Samsung, and Taiwanese industry was mostly composed of many small firms, independent of one another and competing fiercely. Up to 1963, most industries were oriented to the domestic market; a few low-skill industries had exports in this early period, notably in the sector of textiles, and in the manufacture of electric fans and fluorescent tubes. A key sector in the 1960s then became plastics and plastic products, together with electronics. As earlier in Japan, these industries were built up with little investment in research and development, using established technologies which were acquired or used directly by foreign firms in Taiwan.

Those industries that developed in Taiwan were especially suited to small firms, and in this respect there is a contrast with South Korea. Small firms were partly the result of a special integration of farming with industry in Taiwan, notably the long build-up of small mills processing rice, sugar, and other newer crops made important in the 1950s, such as cotton, fruit and vegetables. Most of farming family income, in the 1980s, came from non-farm employment, as farmers became micro-industrialists. In the rural areas, many small enterprises were set up, some on the industrial estates, providing all farmers with alternative employment. This kind of development linked well with the light industry orientation, as the factories could be located anywhere (Scitovsky 1990). A dispersed, rural spread is a feature of some Taiwanese industry. The importance of small firms also relates to the immigrant population which came from mainland China after the Second World War, including many entrepreneurs, and to tax breaks which favoured the establishment of small industries in the first five years.

The automobile industry

A single Taiwanese industry may be used to illustrate the differences from South Korea's model of development. This is the car manufacturing industry, which presents perhaps an extreme example, but none the less illustrates a point.

Comparing South Korea and Taiwan, the latter was the first country to start car manufacture, with an assembly plant under licence from Nissan in Japan set up as early as 1958. South Korea followed this in 1962 with another plant under licence from Nissan. The first difference between the two was South Korea's total prohibition on imported cars, compared with Taiwan's import tariff which allowed some imports. In the late 1960s, both South Korea and Taiwan allowed other competing firms to establish assembly plants in a similar fashion. By 1970, both countries were in a similar position; they each had a domestic market of under 15,000 units, so that the domestic markets would never prove a long-term commercial proposition (Chu 1994). In the 1970s, South Korea set up a special long-term plan for the car industry, designating or targeting specific firms to become the producers of vehicles and limiting the entry of others. It gave several different kinds of aid to exporters, and pushed production to allow exports from 1977. In Taiwan, by contrast, there was no targeting of firms, and large volumes of imports were also allowed, in some years in the 1970s making up 40 per cent of the market. Competition was strong, but this restricted the ambitions of firms that feared the entry of more competitors undermining their own position.

In the 1980s, the South Korean industry achieved nationalization, as the firms, in joint ventures with foreign firms (Asia with Fiat, Hyundai with Ford, Kia with Mazda) took over the dominant control of the Korean operations. In Taiwan, no such nationalization took place, and the industry failed to move forward as in South Korea, remaining at the level of assembly of imports. Tariffs on imports have been a political issue from the 1980s, with consumers asking for reductions, and this has made for reluctance on the part of large MNCs to become involved. South Korea maintained its prohibition on imports up to 1986, and then gradually opened the market to imports, although these are still a tiny proportion of the market.

The South Korean path is more interventionist, and has been more successful in the medium term. It has been possible partly because of the lower level of democracy in the country, and the lack of any opposition to strong policies.

Spatial structure

There was no precise regional strategy, but the concentration of all development was in the west of Taiwan where the densely peopled rice lands and the cities are located. Since 1963 there have been over 70 industrial estates developed in this rural area. Some were intended to use local resources. Others had the main purpose of checking out-migration to the three major urban concentrations: Taipei with its port of Keelung in the north, Taichung in the centre west, and Tainan-Kaohsiung in the south. Large export processing zones were set up, two near Kaohsiung where most heavy industry such as shipbuilding was located, and one in the Taichung region. The main centre of light industry is Taipei. The whole island was provided with adequate infrastructure, including adequate electric power supplies in all areas, roads, airports, and urban services. The success of the industrial estates is disputed, since other rural areas without these estates also did well in the building industry. In another parallel with South Korea, there was some decentralization from the main centres, but it tended to go only to nearby satellite centres. In the whole development period, there has been a more than proportional growth of manufacturing industry in the main city, Taipei. But all the big cities, Taichung, Tainan, Kaohsiung and Keelung, have had rapid growth of industry in their suburban areas.

For the future, evolution of the economy towards high technology and towards services is already happening, as rapid economic development means that cheap labour is no longer available. This means that the factories in sectors like textiles, or any labour-intensive industry, are moved out of the island by Taiwanese firms, most importantly into mainland China. This trend seems likely to continue or even accelerate, with the construction of science parks and office development in big towns.

Indonesia: following the leaders?

A brief survey of this country casts further light on the overall process in east Asia. It might be thought that Indonesia presents a case challenging the model of resource scarcity, concentrated development in the absence of regional problems, and rapid economic expansion. Here there is a country of a different scale and great diversity. Its area is some 2 million km², stretched over 5000 km and broken into over 6000 inhabited islands, poorly linked to one another. The five main territories are Kalimantan (Indonesian Borneo), Sumatra, Java, Sulawesi (Celebes) and Irian Jaya (western New Guinea). It has substantial natural resources of oil and other minerals and of forests, and a large agricultural output. It has great ethnic diversity, with over 300 ethnic groups speaking different languages, and political problems with some of the minorities inhabiting some islands. Despite these features, it has achieved rapid national economic development in recent times, including the opening up to a modern economy of some of the outer provinces where there had previously been a traditional subsistence economy.

Indonesia's history since independence in 1945 is conveniently divided into two periods. In the first, up to 1966, there was strong nationalism under the presidency of Sukarno. This included a separation of the country from Western influences and from most developmental impulses, through hostility to the main Western countries and withdrawal from the United Nations. What little industry arose was protected production of consumer goods in and around the capital Jakarta. Most of the country was either occupied only by shifting agriculturalists (swidden farmers) or, in Java, by rice farmers, an intensive farming system but not one leading to development. Instead, to accommodate more families on the same land and under the same technology, there was the process of agricultural involution, classically described by Clifford Geertz (1963). Rather than producing more, the small farmers shared their work in the rice fields amongst more and more members of the family, and also shared the product amongst the family so as to cover even the poorest and those least able to manage. Involution, in this sense, meant an increasing complication of division of labour, of work responsibilities, and of rights to the produce of the land, with an inexorably growing populace.

After 1966, there were major changes; under a new president, Soeharto, agriculture was modernized to increase food production to self-sufficiency levels by the mid-1980s, using high-yield crop varieties on medium-sized farms. In the industrial sphere, private expansion of conglomerates like the Korean *chaebol* (family-owned and diversified in products), rather than small firms like those of Taiwan, was rapid and continuous. Industrial expansion to produce for exports, following the NIC model, has only happened since 1985, but it has been spectacularly rapid. Up to the 1980s, the only industries were those making small consumer items, and resource-processing industries like the manufacture of plywood. Now the industrial spread of products has become very wide, and there

149

are exports of garments, textiles, electronic goods, petrochemicals, synthetic fibres, motorcycles, as well as the older established plywood, oil and gas, and cement. In the 1970s there was some tendency to expand the state sector with a new steel industry and fertilizer and cement works, as oil revenues gave the opportunity for state intervention. Indonesia remains heavily dominated by state-led industries, but the private sector is expanding and privatization is now in progress. Most of the industry is concentrated in Java, and the large industrial complexes are located around Jakarta in the west and Surabaya in the east.

Rapid expansion of the population, a major problem under Sukarno, was checked by family planning which was gradually extended over the whole nation through the 1970s. Economic growth, and in particular exports, was further stimulated by the opening up of major raw material sources: oil and gas in Sumatra and Kalimantan, and forestry in many of the islands.

Regionally, Indonesia could be said to have been suffering from the kind of problems experienced in Latin American countries, where huge underdeveloped interior regions tempt populist governments to invest heavily in resource development and thus squander away vital funds for human resource development. Major investments were certainly made in the oil and gas industry, and this industry has consistently accounted for over 70 per cent of exports for the country. This oil and gas, mostly in northern and central Sumatra and on the island of Borneo (Kalimantan), was located in a resource frontier, creating the imbalance seen in Latin America between such frontiers, with poor people but rich in natural resources, and a rich centre with few resources on the island of Java. There were and still are strong regional disparities in incomes. In 1971 (Hill 1991), using an index of Indonesia as 100, the poorest province, East Nusa Tenggara, had a GDP per capita of 48, against Jakarta's 248. In 1983 the figures for these two provinces were 54 versus 276. These figures excluded oil revenues from the calculation. Including them would mean that provinces such as Kalimantan have levels of over 600, increasing the regional differences.

Indonesia thus presents an example of rapid development despite its tendency to resource-led development. But it does not really present a model different from the NICS. Much of the development is in Java and around the great metropolis of Jakarta. This city's population of some 9 million in 1993 looks modest compared with a national population of 190 million, but considering the spatial pattern of scattered islands, it is large. Most industries are highly concentrated here, apart from the resource-based ones. Foreign investments have mostly been made in Java, 21.5 per cent in Jakarta and 21.6 per cent in the rest of Java in the period 1967–85. The benefits from industrial development are also concentrated, as shown by the interregional differences in income levels. Oil extraction did indeed present a distraction in the 1970s, which very nearly put development off course, but it was brought back by remedial measures, including an opening of the economy to competition and the promotion of exports to solve the debt crises of the 1980s.

One aspect not tackled with respect to the other NICs may be raised here. This is the problem of ecology and conservation. Because this is a big country, the possibilities for resource degradation are larger. There are also special features, notably the very heavy rainfall, averaging 2650 mm per annum in Java, combined with steep slopes and often loose volcanic soils (Hardjono 1994). Some of the soil erosion problem relates to deforestation, linked to the major forest products industry. In the Lesser Sunda Islands (E. Nusa Tenggara), 59 per cent of the forest land is degraded, and 30 per cent of other classes of land. Some forest destruction is due to shifting farmers, expanding their cultivation into higher lands for food crops, and farmers invading old plantations (crops such as tea gave good soil protection) to plant vegetables. Problems of nature conservation, as in the large Latin American countries, remain as secondary interests among societies where these issues have not been debated and where democratic opinion has not been fully aired.

Macro-scale geographic effects

Development impulses operate at many scales. The largest is the global scale, but below this, and above the national scale, there are some that can be described as macroregional. East Asia provides an example of this kind of phenomenon, where one country operates as the leader, and from it there are linkages to a number of neighbours. For Asia's Pacific Rim, the leader, since the Second World War, has been Japan. This country's economic growth rate still exceeds that of western Europe or the USA, but it is slowing. From the period 1970–78 to the period 1985–94, the growth rate has been more than halved, from 7.8 per cent to 3.2 per cent, and it will probably continue to decline as Japan becomes a mature post-industrial society. These figures should be compared with those for the NICs in Table 8.1, showing a rapid increase through the 1980s.

For Japan the growth pains of rapid expansion occurred in the 1950s and 1960s, when the most rapid economic growth took place. Huge rural–urban migration occurred and interregional differences in income were at their highest levels. Since then, social concerns have broadened out to include regional development, pushing new industries away from the main cities, with an increase in the supply and quality of housing, and the improvement of physical infrastructure such as roads, sewers and city parks, on all of which measures Japan lies well below Western countries (Abe 1996).

But around Japan, other countries of the Pacific Rim have captured the growth syndrome in similar fashion. The most outstanding are the East Asian Tigers (Hong Kong, Singapore, Taiwan and South Korea), all of which have experienced, over the 1985–94 period, rates of per capita GDP growth comparable to Japan in the 1970s. Behind them in the train of development lie a third tier of countries: Malaysia, Thailand and Indonesia. Behind these are yet more countries with an uncertain future but potential for growth: Vietnam, the province of Guangdong (Canton) in mainland China, and more dubiously, the countries of the Philippines and Myanmar (Burma).

It is worth noting two different kinds of mechanism for transmission of the development impulses. These are not always captured by the usual analyses of economists, which deal with government policies or specific industrial sectors and firms. First, there are direct linkages from Japan and from the second tier, down through this group of countries, via trade and investment. Japan has strong trade links to the macroregion, drawing on Indonesia especially for raw materials such as oil and liquefied natural gas, and investing in branch factories in South Korea and Taiwan. Further down what has been called the East Asian escalator, Taiwan is a key player. It was, for example, the largest investor country in Malaysia and Indonesia in 1991 (Government Information Office of Taiwan 1993).

A second level of linkage is through cultural affinities. There is no need to refer to mystical relations or the inheritance of Confucian ideology, sometimes used as a reason for the recent emergence of the region, in order to find the cultural link. Direct migration has brought ethnic Chinese people, many of them merchants and entrepreneurs, to all parts of the region. There are calculated to be 34 million Chinese living overseas, of which 88 per cent are in Hong Kong, Indonesia, Thailand, Malaysia and Singapore. These people are largely from the coastal provinces of China, and from Taiwan itself, which is included as part of China in this reckoning. From Taiwan a specific committee, the Overseas Chinese Affairs Commission (OCAC), looks after these people's interests, working like the British Council in promoting cultural links, language and education, as well as helping to establish or maintain business links. It has promoted investment tours of the ASEAN (Association of South East Asian Nations) countries which bring heavy investment into these countries. In the reverse direction, there has been heavy investment by the ethnic Chinese back into Taiwan during the 1970s and 1980s, and more recently into mainland China directly.

To these elements we may add the East Asian forms of association. In this respect, the West is perhaps the exception, in having made economic life quite separate from social life, so that all economic transactions are made at arm's length. In east Asia, there would seem to be an extension of family life into the economy, which reduces transactional costs because of the ability to rely on trusted employees and business associates (Fukuyama 1995). Fukuyama's interpretation of this phenomenon is as something that lies between the family and the state. A standard view of societies is as individualistic or communitarian, but between these, there are many, including the most successful such as Japan and the USA, that have a strong private social organization. This "social capital" in the form of societies and more informal groupings is put to good use in the conduct of business in such countries, and increases the speed and efficiency of work in larger organizations.

The timing of intervention, concentrated development and the environment

It is worthwhile putting some of the observations in this chapter and the two previous ones into perspective. Three areas will be concentrated on here. The first concerns state policies regarding developmental issues. These may be divided into national and regional policies.

State intervention in the course of development is currently challenged by the neo-liberal school, which instead claims that markets and competition must be allowed free rein. On the other hand, there are those who make the reverse claim that exposure to markets is what has caused the problems for many poor countries. The evidence from the successful countries is that some intervention is desirable. In the cases of South Korea and Indonesia, this intervention continues to the present day, limiting imports of goods that the country wishes to produce itself, and the state shows itself to be a powerful cultivator of new industries under favourable circumstances. This is the case made by Corbridge (1986). For the case of the Soviet Union, the continued state intervention was obviously a burden to the development of Russia (although possibly not to some satellite states.)

A case that might seem to deny the positive role of the state is that of Taiwan, but here again the state has had a role. This was in the earliest stage, when in the 1950s agriculture was totally modernized through the agrarian reform programme, which also got rid of the old elite groups and produced a more equitable distribution of wealth. Thus farming was able to contribute to development of the economy, and the population lost its old social and economic structure. Before the 1950s, state-led (i.e. Japanese) improvements in infrastructure had also been made which provided a good platform for later change.

Another comparable case, which has not previously been discussed in this chapter but which is commonly cited, is that of Chile. In this country, it has been the accepted wisdom that its startling development since the 1970s is due to free-market economics. In reality this is only half-true; in the 1960s under Frei, and then under Allende, agrarian reform took place and the state took control of industry on a massive scale (Martinez & Diaz 1996). Only later were free-market principles followed, building on a new socio-economic base that had destroyed old elites and vested interests.

Another matter is regional policy. This is a policy element for developed countries, like France, Spain or Britain, rather than of those undergoing rapid development. Taiwan, Indonesia and South Korea have had relatively weak policies for regions, and major regional changes have occurred despite them rather than because of them. This might seem to be a fairly obvious conclusion, but it does run counter to the ideas set out originally by Friedmann, who described the bell shape of regional disparity and postulated that regional policy was most appropriate for the intermediate phases when disparity was greatest. There are good reasons why this should be so. Although regional disparities in income and welfare may be greatest for the fast-developing nations, the level of public awareness is perhaps not so high. We have no measures of this awareness, but it may be surmised that in countries that have only recently attained democratic government, awareness of issues high on the political agendas will be poorer than in countries with a long democratic history. Policy to help problem regions becomes important only when those regions, or the politicians representing them, make public protests that must be heard. In addition, people's concern for equity is probably heightened by some of the education and socializing processes of development. Apart from these considerations, it may be economically the most appropriate policy to do nothing, or do the minimum, in terms of regional policy for poor countries. Here it is worth recalling the tremendous migration waves leaving the countryside and reaching cities like Seoul or Taipei, or even Barcelona and Madrid in Spain in the 1960s. It is hard to imagine regional policies that would keep the rural masses in place and employed in these countries, and stem urban growth in the metropolises. Nor could the investment money be expected to be available in these countries. In summary, while national intervention for development seems most appropriate at an early stage, intervention for regions becomes more relevant at a later stage when there are heightened levels of awareness about regions and special groups.

The second major conclusion to be noted is that the recent success stories in development have been spatially concentrated. The cases of Taiwan and South Korea could be amplified by reference to Hong Kong and Singapore, or to the earlier case of Japan itself. Within Latin America, where development has been a much weaker process through the following of different policies, the exceptions have been individual regions where concentrated urban industrial development has been possible. Thus, for example, the region of São Paulo in Brazil, or the US borderlands of Mexico, constitute regions of exceptional dynamism within large countries that have made only halting progress over the last few decades. On what grounds can it be argued that concentrated development should be successful? The benefits of concentration for industry are well known: the linkages between firms, vertical and horizontal, forming external economies for firms, the economies of scale, the size of markets. What needs to be explained is why having extra territory seems to be a negative feature. One suggestion by economists is that focusing a whole national economy on a primary export has negative effects on the rest. This concept is called "Dutch disease", having been identified for Holland in connection with its natural gas industry in the 1960s, which boosted all industries related to gas and raised costs, especially labour costs, across the nation. This meant that Holland's traditional export goods became too expensive and there were economic problems as a result. The connection between Dutch disease and concentrated development is that primary resources tend to be widely spread, so that concentration is not possible. This kind of explanation may work for economies like those of Nigeria or Venezuela, dominated by oil, but it seems untenable for more complex situations like Brazil or China, with many resources. Auty (1994) gives another version of this kind of argument, which he terms the "resource curse" thesis, claiming that size, whether of physical resources or of population, and thus market size, is a problem for the developing country. Looking at several East Asian NICs, and also at Brazil, China, India and Mexico, he finds that the latter countries stayed too long with autarchic policies of the ISI type and delayed their development because of this. Large reserves of minerals such as oil in Mexico, and iron ore in Brazil, tempted the governments of these countries to try to bring in huge and expensive heavy industries. India and China were not resource-blessed, but also tried to achieve ISI autarchy, economic independence, in the belief that their great size of population justified all industries. They continued with such policies long after they had been proven uneconomic.

More acceptable to students of the countries concerned may be an explanation of failure of unconcentrated development, which starts with inward rather than outward orientation, but explains the inward orientation both through size of resources/population, and through the social and political processes of the countries. Heavy industries were indeed wrongly emphasized in Brazil and India, not simply in the belief that resources existed, or that markets existed, for the products, but from geopolitical stances in the countries concerned. In Argentina, as shown earlier, and also in India, China, Brazil and Mexico, it was a matter of considerable national pride to have a strong iron and steel industry, at any cost. Military governments also in Latin America saw a strategic need to implant iron and steel industry to ensure supplies to the domestic military machine. In most of the countries concerned, industries such as iron and steel could be portrayed as regional policy (Las Truchas on the Mexican Pacific coast; backyard steelworks in rural China; iron ore and steel works in Minas Gerais, Rio and São Paulo states in Brazil). The failure in these countries is the failure to adopt as a driving element the concentrated industries of the big cities, and overinvestment, for a variety of reasons, in industries located in the peripheral regions.

Environmental costs

Less developed countries with a programme of rapid development face one set of problems, those of environmental conservation and management, in a particularly acute form. Programmes of economic development in such countries are likely to start or to amplify existing processes of soil erosion, forest destruction, and pollution of the water and air, with little control and with dire consequences for the future of these countries. This is true at the national level and has been commented on in many countries. What is sometimes not appreciated is that at the regional level, the effect of no control is still more acute, as some regions depend entirely on one product or resource.

It is worth noting why the LDCs should have a particular liability to suffer from natural resource destruction and deterioration. In the first place, development initiatives generally come to these countries from the outside, from another country, from the world markets for particular products, and through the action of MNCs. Because the thrust of development is externally driven there is little interest

155

in environmental protection, which will not help the outside company or country in any direct way. This may be regarded as an extension of the dependency argument, in that lack of care for the environment is part of the exploitation theorized in dependency writings. In the case of MNCs, even if they have some interest in protecting the supplies of the resource they are themselves using, such as quebracho wood which grows in Paraguay and has long been used as a source of tannin, their timescale of interest is relatively short. It is most unlikely to include the 200 years necessary to regrow the quebracho, which is a slow-growing tree in dry tropical forest. In the case of fossil resources such as oil, they do not have any interest in the diversification of the local economy away from oil after the reserves have been exhausted.

Local indifference and the case of China

Control by the MNCs is not an adequate explanation in many important cases, however. Mainland China has been recognized for some time as having highly polluting industries, based on coal being burnt in an inefficient manner, and this has been at a time when that country tried to sever virtually all contacts with the West. But even when China has sought to move forward from coal, it has had little concern for the environment. The classic case is that of the Three Gorges Dam on the Yangtse; this river, the third largest in the world, formerly had some 62 smaller dams on its middle section, which were all swept away in a powerful flood of 1975. This gives little cause for rejoicing at the present plan to build one huge dam near Yichang, for the purpose of energy generation and to control floods.

The new dam is on a geological fault, and will produce a reservoir nearly 400 miles long, submerging much farmland, nearly 800 villages, and many historical sites of interest. This dam is likely to be built (work has begun) despite opposition from international agencies such as the World Bank and the United Nations. The dam is built in a country where conservation issues and the environment have relatively low value, at least to the leaders, and where the voice of opposition coming from many Chinese in the country is stifled by a dictatorial government.

In many LDCs in Africa and Latin America, it would appear that local governments have little interest in any case in environmental protection or sustainability of development. Where there are non-representative governments, whether military or civilian, there is often a determination to promote development with a minimum of extra costs, such as those of maintaining the environment. A somewhat similar case is that where one sector has been taxed to benefit another, and the workers in that sector are thus unable to invest at all in conservation. In Argentina, throughout the Peron era of 1945–55, agriculture was burdened with export taxes and revenues were invested in manufacturing industry support. In South Korea the same thing happened in the 1960s, not through export taxes, but through the control of internal markets for grain and compulsory purchase at low prices (Edwards 1992). Under the pressure of low prices, peasant farmers were forced into more intensive cultivation with less rotation of crops and less rest for the soil, resulting in soil erosion.

Other explanations

Another reason for the environmental problem is that LDCs have insufficient local capital available to be able to contemplate any kind of investment in protection and conservation. Controlled development of the Amu Darya and Syr Darya rivers irrigation programme in Central Asia would have required additional expenditure on diversion of north-flowing rivers such as the Ob to replenish the Aral Sea. Such expenditure was not within the capacity of central Asia republics, and in the event, not seen as possible for the Soviet Union as a whole when compared with other possible schemes for investment.

Another set of phenomena explaining the special problems of the LDCs is that surrounding the knowledge systems involved. Sudden invasion of a region for growing a new, or even an old, crop, commonly fails to take up local knowledge about the environment from the indigenous population, people who may have developed sustainable management systems over centuries. An extreme case is that of interior Brazil, currently being opened up by loggers, ranchers and small farmers from outside the region. They commonly have little experience of rainforest environments, and as a result, set up destructive land-use systems. In the Amazonia region, the Amerindian natives have long evolved forest-using systems of shifting agriculture and silviculture that do not destroy (Browder 1989), and the mixed-race peoples who gather forest products such as rubber and Brazil nuts also do so on a sustainable basis. These local systems are ignored by the invaders. Up to the present, the record is more of elimination of the local population or its removal to other regions, rather than incorporation into the development process.

In other regions, such as the Central Andean mountains of Ecuador, there is a peasant population that has lost its traditional links with the environment and much traditional knowledge. Pressed for production on tiny farms, it overexploits these by growing maize as a monoculture on steep, unprotected slopes, causing soil erosion and flooding or sedimentation problems downstream on the rivers. In this kind of case, now common in India and Africa, the local population has lost traditional knowledge but does not have the advantage of modern scientific knowledge to compensate.

All these problems exacerbating the overall environmental pressure in LDCs are further aggravated in specific regions. At the national level, there is usually some diversity in production systems. In an individual region, only one resource, whether it be the forests in upland Indonesia or cotton in Central Asia, is the main focus. This imposes a direct strain, but it also means that when the resource is eventually exhausted, there are no other arms of the economy to turn to in order to help restore the regional economy, or to pay for restoration of the environment.

Particularly in the thinly peopled peripheral regions of Latin American countries, the decline of a single regional resource, such as copper or tin mines in the Peruvian or Bolivian Andes, or sugar plantations in northern Argentina, leaves the region economically fragile and may cause a segment of the population to emigrate permanently from the region.

The response

In developed countries, there is a long history of reponses to threats to the environment, coming from different sources. In Victorian England, William Morris sought to re-engage industrial humankind with the environment, claiming that a whole life could not be lived in the abstract environment of the city. He believed that people should always be engaged in direct contact with nature and that their work should include making physical objects: that in following pre-industrial lifestyles, we should be craftspeople, which would immediately bring an awareness of environmental values.

In recent times, movements such as that for Development from Below (see Ch. 2) have taken up comparable themes, seeking a rounded, internally organized development process, which would also be conservative of natural resources for the locality or region because it was organized by the local people. The problem remains that this kind of grassroots movement remains very much a theoretical, or if existing, then an artificial and supported kind of movement. An example was given by Smith (1992) of Indian development projects. On the one side, he noted the Narmada Valley programme in the northern Deccan. This top-down plan involves the building of 30 large dams and thousands of small ones, and the irrigation of 20,000 km² of land. It would be paid for by the World Bank, at a cost of $40 billion. On the other side, a tiny grassroots project near Delhi is that of the village of Dhanawas (population 300), where sustainable development projects, worked out in cooperation with locals, include planting trees on poor land, producing biogas for fuel for cooking, and use of solar-power cookers, to limit the traditional use of dung and allow it to go back on to the fields. There is no record that the Dhanawas experiment, supported by a large industrial company, is extending into the countryside generally. Large projects with outside finance remain the most likely.

CHAPTER 9
Regions in their own right

One part of the geographer's traditional lore, set aside in the quantitative revolution but reverted to again in the 1990s post-modern concern for interpretations of places and for research into localities, has always been focused on the special and separate features of each place or region. It is appropriate to examine, therefore, to what extent localities or regions need or will pursue separate development courses, and whether these justify a separate planning exercise.

With some exceptions in the consideration of specific countries and local problems, in this book we have treated regions as parts of a whole, with the idea of national development being distributed spatially rather than regionally, in the sense of there being homogeneous spaces rather than individual regions to concern ourselves with. Geographers have long argued over the issue of the general or specific character of spaces or regions, and most of the argument has gone in favour of spaces, because if all areas can be assumed to be the same, in some basic sense, then they can be dealt with in a more mathematical way, and the existence of any awkward unique features may be overlooked. But in fact many features of regions are unique, within one country, and even where the features are not exactly unique, the way that they combine into a whole, as argued eloquently by Richard Hartshorne 60 years ago (Hartshorne 1939), is itself unique for each region. Regional development theory itself was criticized in the 1980s, for treating of space rather than specific regions (Gore 1984).

From the time of early modern industrialization, there have been reiterated schemes for separate kinds of development, often calling for self-sufficiency and separation (Weaver 1984). At times the voices have been muted, but following the economism of the 1960s, the 1970s and 1980s saw a strong current in the social sciences which sought to identify another way for regions to achieve their own development, not forced by the centre but using their own human skills, physical resources and local initiatives. This often came from the planners themselves, calling for new organizations of development, such as territorial development, rural development, Development from Below, and endogenous development. In the 1990s there is another current, more radical because it is not attached to any formal planning structure, which is sometimes identified as anti-development, and sometimes as "another development" (see Ch. 2). A key point throughout this discourse on development has been the lack of separation between the course of development, and the plans or policies to actually produce development. Much of the writing was in terms of some ideal goals, and reference to real-life processes was limited. This chapter starts by asking whether there is a factual basis of separation which gives a lead for separate planning, to look next at examples of locally based development, and then to examine the policies advocated.

Regions, it may be noted, can be distinctive in terms of their physical structures, whether we consider physical resources for development, or the simple geographical location in relation to other locations, or some set of environmental problems stemming from topography, climate, soils or vegetation. They may also have human uniqueness, in terms of a long cultural history or ethnic difference. Any single one or a combination of these features may constitute a basis for organizing developmental programmes separately.

The chapter also looks at political power. On the one hand, local, separately organized and separately controlled development depends on either the lack of central power, or a central power willing to delegate and decentralize. On the other, the pressure for regionally organized programmes may stem from and rely on political activism, even when the local economy is not particularly different from that of the centre.

Regional distinctiveness

The following section tries to give an account of the kinds of factors that call for locally based development actions. Most of the ways discussed are policies rather than theories, or desired policies, highlighting the special needs of the poor, of the rural, and of particular minority groups. As Taylor & Mackenzie noted (1992: 234), this is making a theory out of policy.

Some of the factors providing a region with its identity are those of the natural environment, which shows infinite variability in topography, climate and mineral resources. Some regions are mining regions, others are characterized by their forest or water resources, and yet others have a special feature in their remoteness because of their location in relation to the industrial centre of the country or to topographic accidents such as high mountains and rivers. Other factors are social, cultural and political.

Regional environmental problems

Distinctive environments, those that are unique within a nation, are often a key feature demanding separate attention in the course of any developmental programme. In Russia and the Ukraine, soil erosion in the "Black Soil" (Chernozem) region is one such a key feature. Further east, the Aral Sea region's desiccation is an important issue which affects the region around the Sea and the rivers that feed it (see Ch. 7). Other problems that distinguish individual regions are: the permafrost of Siberia and northern European Russia; nuclear waste in the Barents Sea, from dumped reactors of submarines and rotting contaminated vessels; and oil pollution in the older oilfields of western Siberia. But consideration of these latter for nomination as special target regions raises immediately the question of whether central government should be the administrator of regional policies, since it was the creator of such problems.

Another set of related problems attaches to regions that are isolated or remote from major centres. All rural regions, to some extent, "suffer" from this characteristic, in comparison with towns and cities linked into a national network. But some regions suffer in particular ways, because of a combination of location and environmental difficulties. Thus mountain regions, such as the Andes in South America, the Urals in Russia, or the Alps in Europe, have a collection of special problems due to their location. Transportation and communications may be of particular difficulty in such regions.

Recent recognition of the problem has occurred in Europe, where the European Union (EU) has designated a new category of remote region under Objective 6 of its regional policy, involving the development of sparsely populated regions in northern Finland and Sweden. All the other Objectives from 1 to 5, for what are called the structural funds of the EU, were defined in terms of economic and social variables directly – unemployment levels, income levels, lagging industries or backward agricultural systems.

In Britain and Norway, similarly, separate regional agencies were created some time ago to cater for the remote northern regions. These regions (the Highlands and Islands in Scotland, and Nordmark in Norway) are treated separately on the basis of their location and qualitatively different problems, rather than lower living standards or employment levels.

Separation may also be justified in the case of regions with a special resource endowment. For regions that are otherwise poor, with few inhabitants and few sources of employment, the presence of a single large resource presents problems because of the tendency of the nation to use the resource in a single massive boom, and thereafter to abandon the region to its fate. To take a country in South America, Chile has always had great physical resources, mostly located in remote regions where there are no alternative economic resources at all. Thus the northern desert region of the Atacama saw booms based first on silver, in the colonial period, then nitrates in the second half of the nineteenth century, and in the twentieth century, copper from mines which are mostly high in the Andes. In southern Chile, a resource-based development was that of steel at Concepcion, using local coal and iron ore from the north. This has become uneconomic, and more important today is the forest cover of this region. This is being rapidly exploited, to be replaced in this case, but only by lower-grade softwoods.

Such situations of boom and bust are also found in countries such as the UK. Within Scotland, for example, the Western Highlands experienced several minibooms, connected with, in the eighteenth century, the exploitation of oak forest for charcoal and ironmaking; in the early nineteenth century, the collection of kelp for its ash used in soap and glass-making; and in the later twentieth century, a boom based on the use of hydroelectric power for aluminium, and local planted forest for pulp and paper-making, both of which have now been severely restricted because of competition from other countries.

Summarizing the above, it is perhaps a fundamental feature of remote regions that they have few alternatives to the single product economy, and so the seesaw conditions of boom followed by decline are inevitable. This certainly justifies them in having special policies, from whatever source. Whether it also implies separate policy-making, at the local or regional level, is less clear.

Human groups and separate interests

Another kind of basis for separate organization of developmental programmes lies in the human groups occupying regions, which may have distinct and different aims for their development. In the early chapters, the case that development has many dimensions other than the economic was reviewed, and emphasis put on the social aspects. The point about economic development is that it has tended to be unidimensional itself (i.e. movement towards universal consumption of standardized goods and services). If development is understood in this way, there is no good case for separate development as there is a common aim for all people. Once separate aims are conceded, though, the case becomes clearer. Stöhr & Tödtling (1978) examined this matter in some detail, building on the ideas of sociologists and psychologists, and discussed three kinds of dimension in the pursuit of human happiness, apart from the purely personal ones. These three dimensions they described in terms of concepts of "having" (broadly speaking the economic dimension, material possessions), of "being" (status or prestige accorded to the person within their society), and "loving" (the friendship or goodwill between members of any human group). From this analysis, and from the discussions in Chapters 1 and 2, it becomes apparent that the social dimension (which may be taken to include the being and loving dimensions of Stöhr & Tödtling) has great importance.

From this it is possible to go on to claim that social awareness, strength of group feeling and the rewards it brings, will be highest where there is some autonomy of the local society, including the organization of development efforts. Strongly organized societies and groups have emerged in many parts of the world and in different contexts, but they are evidently favoured where top-down controls are limited and not too severe. This line of thought obviously inspired Stöhr's later espousal of Development from Below (1981). At the extreme, in dictatorial environments, such as that imposed in the Soviet Union over most of this century, the only organizations of society were those endorsed by the state, and people suffered the loss of alternative expressions of social coherence. In less extreme situations such as in Latin American dictatorships, the family or extended family has remained as an important social unit, but one with obvious limitations because of its small size and scope. But in this latter region, the breakdown of authoritarian government in the 1970s and 1980s, especially in countries such as Argentina and Chile, has led not just back to the family, but also forward to the growth of a variety of institutions, usually labelled non-governmental organizations (NGOs), to fill the vacuum left by the state (Reilly 1995).

In much of the developed world, there has been freedom for many other social organizations, which represent economic groupings such as firms, institutions such as universities, and those with ethnic differences or long cultural histories expressed through a distinct language. These are structures that give the opportunity for the interaction and expression of social coherence, separate from the economic, and often very disparate between themselves.

At the highest level, these cultural distinctions are expressed through a political unit, and many subnational political divisions represent an early economic, social and cultural break. Politically based differences between regions form a solid basis for regional differentiation and regional planning. Within Britain, for example, Scotland, Northern Ireland and Wales do form logical units for regional planning, and to a lesser extent, and at a lower scale, so do the counties within each country. The UK in fact provides an example of demands for separate organization of life, including development policies – demands that are articulated at the level of the historical nationalities of Scotland, Ireland and Wales.

Environmental concerns

In between the physical distinctiveness arguments and those for human organization summarized above, there are a number of points that are currently rated as important, connecting humankind and environment. A central concern is for "sustainable development", a term that has been defined in various ways, but which is most frequently understood in the manner of the Brundtland Report (World Commission on Environment and Development 1987) as "development that meets the needs of the present generation without compromising the ability of future generations to meet their own needs ...". To achieve this kind of development, or rather the conscience that will control the levels and rate of use of resources, requires a measure of local initiative and power to administer programmes (Taylor, in Taylor & Mackenzie 1992: 214–58). Taylor also reports the lack of national concern for managing the environment in Africa, so that the only hope must be with local bodies. Another related point is that given the great diversity of environments, and the long history of their use by locals, it is necessary to rely on local and indigenous knowledge systems to husband the local resources. Scientific knowledge has often been proved to be inadequate for the management of, for example, the forestry sector in tropical countries, and has worked in ignorance of the effects of some of its proposals.

Another kind of "environmental" view justifying the local, has been that

163

regarding "territory". It may be accepted that for most people there is an emotional attachment to particular areas or regions, which have been their homes for a long time, their childhood lands, or perhaps even the lands of their forebears. Maintenance of the physical environment in its traditional form, avoidance of any destruction, and concern for the wellbeing of the residents of these local areas, are components of the territoriality that is found not only in the LDCs but also widely in developed countries, where the fierce defence of landscape relates to the same feelings. Most Western countries have agencies to conserve or preserve elements of both the built and the natural landscape, with strong support from people concerned for the visible symbols of historical identity.

Ways towards separate regional development

How can a locally based development be planned and executed? John Friedmann's 1970s ideas, and in Europe, those of Walter Stöhr, set in motion a variety of critiques of the standard top-down style of development, and a variety of other related approaches sought to identify how separate development styles could be engendered by moving away from the top-down direction. The search for ways of promoting and organizing local or regional development from the grass-roots gave rise to a flourishing literature on aspects of "Development from Below", summarized in the book edited by Stöhr & Taylor (1981). One version of it had emerged from the International Labour Organization and the United Nations, and concerned the supply of basic needs. If regional development could be directed simply at the supply of housing, clothing, food and basic services for the poorest groups in each region, then a new style of development would have been created.

This was regarded as Development from Below presumably because locally based development could be expected to attend to local requirements first. This approach was full of the best intentions, but was agnostic as to the key matter of how it might be achieved. Like the other versions now mentioned, the concern was for policy, which was then presented as a kind of theory.

Another version was agropolitan development, as espoused by Friedmann and others for the East Asia area (Lo & Salih 1981). This concept was expressed as a series of desiderata: development should involve the local community at all levels and all points in the planning process; products of the region should be, as much as possible, from resources of the region itself, using the regional technology and other human resources; production should be for the local market; all barriers to production, such as the unequal division of the land, should be resolved by agrarian reform. Again it may be seen that the aims are laudable, but the mechanism for action difficult to define. There was a specific spatial structure for agropolitan development – the regions were designated as having up to 150,000 people in them, with a central city having an initial population of between 10,000 and 25,000. These dimensions would ensure that a true democracy would reign, since the small size of the region would allow all its residents to visit the centre and take part in its governance, rather like the Greek *polis.*

Yet other versions emerged in the same period, most notably ecodevelopment, and then ethnodevelopment (Hettne 1990). Ecodevelopment (see also Ch. 2) had to be locally based, because it was concerned to conserve local resources and not to exceed their capacity, which was always associated with outsiders coming in to exploit these resources. Ecodevelopment has had the greatest impact on theory of all types of Development from Below, because it has been this branch that was converted into "sustainable development", once global concern over global resources was translated into discussion of possible action and advertised through the Brundtland Report (World Commission on Environment and Development 1987). Ethnodevelopment concerns the identification of specific groups in society with distinctive capacities and knowledge, and their promotion without bringing them into conflict, usually with other dominant groups. Such groups might be racial minorities, or groups with a distinctive, historically identified culture.

The above are all examples of a recognition that some regions have special problems, needing a one-off, individual approach. A simpler version of the approach, allowing that rural regions are different from urban regions, was rural development. This was an early reaction to the fact that regional programmes of aid tended to help the cities and their industries, and actually made problems worse in the countryside. From the mid-1970s, the World Bank (1975) has acknowledged the special problems of rural regions, especially in the Third World, where there is outmigration of the best-qualified individuals to the towns and cities, leaving behind the poor, the less qualified, and the old. Lack of industrial development in such regions, and the lack of social provision, have been constant themes.

Rural areas are clearly not self-contained in the modern world, and rural development as a theme had soon to be modified into integrated rural–urban development. Integrated development of towns with their countryside could establish cash crops in the farmland, and the agro-industries to process these crops in the towns, which acted as market centres. The market centre function could itself be upgraded, since in many such regions a major failing was the lack of transport from farm to market, and the lack of marketing institutions and transport agencies to get farmers' goods into the market. Poor levels of services and welfare in the rural area could be addressed by establishing schools and other facilities in the local towns, together with organizations to ensure that the service would reach out into the countryside.

Not all aspects of rural development were Development from Below. Many of the schemes enacted were funded by central government and international aid, and directed also from the centre, but they did recognize the imbalances created through economic development and tried to encourage more local initiative and control over the process. Much of the writing, such as that by Rondinelli (1985) or by the World Bank (1975), treated the rural regions problem as a generic one, soluble through a suitable settlement policy allied to sectoral moves to generate markets. Rondinelli saw the problem as one of poor countries having no market towns where the processes of exchange of surplus could take place, and creation of these market towns with their markets would initiate the development process. This was putting the cart before the horse – markets and market towns would only develop where a surplus was being generated – but it also overlooked differences between rural regions.

Can we find examples of Development from Below?

A truly locally based development requires local programmes and initiatives to be taken, effectively a "Development from Below". This in turn normally requires some decentralization of political and administrative powers, from a centre to the individual regions or peripheries involved. It also requires the presence in the regions of local organizations prepared to handle the decentralized responsibilities. The case to be made here is that there are usually no local organizations with these levels of competence, so that separate regional development has not happened, especially in the LDCs. It is probably the case, even in developed countries, that initiatives will not be taken without higher authorities and organizations.

Let us take, for example, the case used by John Friedmann with regard to the USA (Friedmann & Weaver 1979), on the most noted river basin regional planning exercise in that country. The Tennessee Valley Authority or TVA, set up in 1933, had as its role in the early years a combined set of purposes, integrated through the river basin itself. This included: the control of soil erosion on the steep slopes of the middle Appalachians, by teaching farmers better techniques of farm management, such as planting maize across slopes rather than up and down them; the building of dams to check floods on the river, as well as to allow better navigation on the river; the use of the electricity generated from installations at the dams, to provide power to the rural communities and isolated farms, allowing labour saving and increased productivity; and the improvement of the infrastructure of the rural communities with better roads and other facilities, so that the bad old subsistence farming methods would be changed.

As originally planned, the TVA programme was acceptable to all, and hailed by the regionalists who had argued for strong local communities, decentralization, and the preservation of traditional regional bonds against the growing centralization of the nation (Weaver 1984: 70–71). It was not claimed as Development from Below, but it was presented (by Friedmann & Weaver 1979) as "territorial development", meaning development undertaken with local or regional concerns dominant, rather than those of the state. In reality, however, it was always a national scheme, within President F. D. Roosevelt's control and outside the control of individual states. It was a plan to promote national development through the then new methods of Keynesian demand stimulation: in other words, pouring public money into massive works, which would employ people, and thus would recirculate capital into the economy at large, giving it the kickstart that would enable further economic growth.

Friedmann and Weaver's thesis was that in the early years, the TVA was an inheritor of true regionalism and a regeneration of local initiative; only in the post-war era, when a need for nuclear power became apparent and agriculture in poor regions was no longer a priority, did the programme change as the TVA was subverted by national concerns, which were not those of the region. Instead of a truly integrated regional agency for multipurpose river basin development, the TVA it became an organizer of mass power production, between nuclear power plants and those already producing hydroelectric power on the Tennessee River and its tributaries. Its other functions and its links to the local communities were lost, and with them the whole "territorial" style of regional development. Functional development (for the nation) replaced territorial development (for the region).

This assumes that the territorial version was always there ready to happen, and that it was merely suppressed by national concerns. In fact, the nation was involved from the beginning in the programme for the Tennessee River, a truly top-down organization, and had the TVA relied on local initiatives, it would never have happened. There was no basis for Development from Below in the region, and the rural communities of the southern Appalachians, often composed mostly of poor sharecroppers, were in desperate need of outside help and unable to apply any kind of endogenous initiative for their own development. The moral of the story is that, in the world's richest country, a scheme for regeneration of a poor region could not be left to the local inhabitants alone. If this was the case for a poor region of the economically most advanced country in the world, it would obviously be still more relevant to regions in poor countries.

Other examples

From the development record there are remarkably few examples of successful and long-term regional development, whether planned or spontaneous, occurring from the grassroots. In addition, as commented on elsewhere (Ch. 2), it appears impossible to predict where these new regions might be. A historic example from Latin America was that of the Antioquia region of Colombia, around the city of Medellin, its economic development based on a large number of independent miners working in the mountains nearby whose small capital was then transferred into the coffee and other industries (Morris 1981). This region also benefited from the relative absence of the more typical social structure in Latin America, of large estate owners dominating economy and society, contrasting with small farmers with no capital.

Modern examples are few and far between. Stöhr (1981) cites some examples of "self-reliant development" from Asia and Latin America, but these have not generally extended to other regions, and most have disappeared in the period since

167

he wrote. In the text by Taylor & Mackenzie (1992), which attempted to follow up the thinking of Stöhr & Taylor (1981), there are further examples, but the message is still not clear. In Africa, from which the later examples are taken, the initiative is sometimes from outside, from central government. This is the case for northwest Ghana, where a credit union established for helping the most impoverished members of society was effectively taken over by the more powerful local elite, and its purpose subverted. This case contrasts with another cited example in southeast Ghana, where a community combined in a number of ways to combat the multiple crises of drought, bush-fires, and numerous refugees from civil war in Nigeria. What Taylor & Mackenzie (1992) demonstrated was the continuing fragility and variability in success of grassroots schemes.

An example of greater scope is that of the Spanish Basque region (Campbell et al. 1977), where an old industrial region around the town of Mondragon built up its industrial might in an innovative cooperation movement. It had long had engineering industries, notably foundries and manufactures of small metal components, existing in the hinterland to Bilbao as main industrial centre. But these had stagnated in the twentieth century and had been hit badly by the Civil War. In the 1940s, a young Jesuit priest, Father José Maria Arizmendi, came to Mondragon, saw the need for training and support to industry, and set up a technical school in the town. The first graduates of this school set up their own industries, but had also imbibed the idea of cooperation, needed to gain market strength in an industrial sector where new products made of many components, economies of scale, and marketing would be needed.

From 1956 producer cooperatives began to be set up in the Basque region, combining the forces of the various small firms, most of which employed less than 50 workers, and thus giving them greater security of employment. On top of the producer cooperatives, a second-level cooperative was set up which afforded technical and financial advice, and concentrated the savings of the worker families. Another was then set up to provide insurance and pension services and other welfare matters. The total number of firms operating under producer cooperatives in the region is now over 150.

This structure has proved remarkably successful, creating new opportunities as new products have been identified, and having the flexibility through its small firms to allow adaptation to new requirements. Democracy at the firm level, with worker directors in rotation and each worker being required to purchase shares in the cooperative, and thus having a financial investment in the company, has ensured good industrial relations, in contrast to much of Spanish industry.

The example is inserted here to show the very special conditions under which this development took place. Commentators have often noted that the failure of Mondragon-type development to spread to other areas reflects special conditions – the special community culture of the Basques, and the existence of many small firms engaged in one sector in the area, each able to benefit from the presence of others, enjoying economies of scale through combination, and external economies in technical and financial advice as well as in marketing the products. These are, economically, approximately the conditions for flexible industrialization. Certainly the development has been without the help of the central state, and does not rely on it now. Compared with the rest of Spain, the Basque region has a stronger social infrastructure, with traditional small-scale communities having powerful internal links. In the Mondragon region, roughly the Deva valley 30 miles south of Bilbao, there was the technical structure of many small cooperating and competing firms, which could build on this social structure.

Central power and the regional response

All the efforts at developing a theory of endogenous development in small regions failed to address, or treated only superficially, some underlying problems regarding the power structure of countries. Power in most poor countries tends to be highly centralized, and decision-making is done by a few people in the capital city. When this is true, the bias of investment and support is towards the interests of the centre, commonly the bureaucracy, manufacturing industries, and the high-quality services provided to the residents of the centre region.

This has been true for the former Third World countries, but also for the Soviet Union and China where command economies were more fated to be centrally controlled than the social market economies of the West, because of the greater need for a state control system to handle all kinds of decisions about supply, demand, prices, etc., in the absence of open markets. Even in countries such as those of Latin America, where the written constitutions are modelled on those of the USA, centralization over the whole Independence period has been the order of the day, with no real power invested in the provinces, and power often invested in a president alone, or a military oligarchy.

Here there is an imbalance between the elaborate theories of geographers and planners for Development from Below and its variants, and the lack of any models for decentralization as a formal undertaking by the state to endow regions with power and resources. On the other hand, this decentralization process has been observed by political scientists for some time now, and there are real world examples of the process.

Decentralization

Decentralization clearly has a variety of different meanings. At a first level, which is termed deconcentration, it may mean simply the removal of some offices or factories to outer regions. Decentralization may involve the splitting-up of decision functions between the regions, so that autonomous offices and head-quarters may exist in various places. In industry, deconcentration might be for a beer manufacturer to set up various new factories in the regions, while maintaining one headquarters for all research and administration. Decentral-ization would involve possibly separate firms within a brewing group, producing their own beer and having control over the process and the products.

In government, deconcentration might mean moving one or two ministries to different regions of the country. Decentralization is, by contrast, the changing of the administrative structure for one or more ministries, so that regional offices are given a greater role. The extreme version of such decentralization is devolution, a wholesale granting of powers to regional or local governments.

There are examples in Europe of the devolution of power to regions, and given that this is the largest form of decentralization, the effects might be expected to be visible in such cases. A well-known version of devolution is that afforded by Spain, where between 1977 and 1978 the highly centralized Francoist state was formally moved to a system of 17 autonomous regions (see Ch. 6). In theory, with the devolution which has taken place in this country, there should be evidence for separate and different paths of regional development in Spain. In reality, there are several reasons that it is difficult to identify a real differentiation of regional paths, here as in other European countries. First, the central state still holds the reins of economic power, with command over economic development and major industries. Secondly, in the key economic areas, what has happened to the regions has been constrained by a lack of economic development in the period since 1978 in most sectors, giving no scope for new initiatives. Thus, Catalonia has been hit by stagnation in its tourism income (already threatened in 1973 but affecting the industry severely in the 1980s) and by crisis in its traditional textiles industry. Valencia has suffered closure of its steel mill at Sagunto and reduction of employment at the Ford car plant at Valencia. The Basque region has had a continuing crisis in its heavy engineering and steel industries through lack of markets for the products, like the rest of western European steel industries, overshadowing any progress made with newer industries such as the light engineering at Mondragon. Agriculture, in regions such as Castille-Leon and Castille-La Mancha, Andalucia and Extremadura, is dominated by changing regulations within the EU and the reduction of support for small farmers which is beginning to happen. Regional development processes are hard to detect when there is no overall development going on. A third set of constraints has been the degree of state ownership of industry. Under Francoist Spain, the economic model was that of the corporatist state, intervening actively to support and control certain activities. The best example of this was the INI, the National Industrial Institute, which was responsible in the mid-1980s for 10 per cent of Spain's GNP and dominating certain industries such as shipbuilding and aluminium. This structure of state control and intervention is only now being dismantled, and separate regional initiatives will take a long time to replace it.

Conclusions: putting scales together

Regions do have a separate identity beyond that of being a unit within a spatial structure, and regional planning to suit each region is thus warranted, although because it is a distinctive enterprise in each region, individual handling of the problem is required. In theory this might be organized either from below or from above. Development from Below does not have a good record as a planning tool. It emerged as a desirable aim in the 1970s period of reaction to previous regional planning, but never entered any country's planning mechanism as a permanent institution. It may be true that powerful developmental impulses emerge in some regions, like that of Mondragon, from local sources, but these are exceptional and cannot be planned or projected. In the LDCs, it would appear that local-based, broadly defined development initiatives are still a rare occurrence, and tend to have only an ephemeral life.

Counterbalancing the arguments for locally based development are the many observable facts about an increasingly global economy. The scale of many important economic activities is far larger than any traditional region. Steel manufacturing, for example, must be planned for on a national scale, or preferably on an international scale for most small countries whose consumption does not match the output of one large mill.

But most discussions of the subject have argued from one extreme or the other in terms of spatial scale. In reality, and in practice in many countries, there is a combination of scales used. While considerations of conservation, urban expansion into rural areas, management of pollution, or the use of physical resources, are discussed in local and regional debate, national and global-scale activities are considered at higher levels. One type of organization does not necessarily negate the other. From what was said earlier about the specificity of problems for individual regions, it is not necessary either for each small region to have its own agencies. Only where there is sufficient objective or felt difference about the aims and interests of the region is a separate organization justified.

Many of the regional problems in development may arise from arguments between the levels, about who should be in control. The forced development of Catalonia's Costa Brava in the 1960s (Ch. 6) reflected Franco's centralist government concerns for national economic development, and Catalonia now seeks a very different kind of tourism development. The forced development of the central Asian cotton fields in Kazakhstan and Uzbekistan (Ch. 7) reflected Moscow's central concern to increase cotton production, against local concerns for conservation of older ways of life. In these cases, the region lost for want of local power, but the reverse can happen. In the UK (Ch. 6), the steel industry was expanded by implementing huge new integrated iron and steel works in Wales and Scotland in 1954, to meet strong regional demands within the national government for investment in heavy industry. In Scotland's case the move was, in the long term, poorly judged, as the Ravenscraig mill had to be closed in the 1990s as an uneconomic site. Massive restructuring of the local economy has been the painful result.

These examples are not particularly about development initiatives from below, but they do illustrate the fact that most development planning can be undertaken at different levels, and that the privileging of one level to the exclusion of others will create problems. They also illustrate another fact: that conflict between different levels of interest is likely to arise in centrally ruled non-democratic states, such as Spain under Franco, and Central Asia under the USSR rule. Room for local initiative to emerge is only likely to be made within liberal democracies. It is also to be noted that most of the LDCs have only a limited kind of democracy, often with single-party rule, so that regional and local initiatives are also difficult within these countries. This makes the current calls for sustainable development, endogenous development, or Development from Below, little more than idealism for these poor countries (Taylor & Mackenzie 1992). Local initiatives must wait until a more favourable political atmosphere has allowed civil society to advance within these countries.

CHAPTER 10

Geographers and development

There are a number of conclusions that may be drawn from this book's review of development issues. Some of these concern the nature of the development process itself, which has always been a bone of contention. Others concern the policies and plans that may be enacted with respect to regions, and the problem of whether intervention of any kind is justified. The enquiry has taken account of observations of the course of development in different countries, and the conclusions draw perhaps more on historical experience than on theory, which has often lacked evidence that it may be translated into practice.

Because this book is written primarily for geographers, consideration is also given to the way in which geography comes into the argument. Different comments about spatial aspects and how geographers might contribute have been made throughout the text, so that the purpose of this final section is to bring together these elements to form a coherent statement on, first, the temporal and spatial patterns observable in development, and secondly, the ways in which geographers might study or work on development themes.

The course of development

One essential message throughout has been that the development process is irregular in the dimensions of time and space. It also varies in its actual content and aims over time and between different human groups, so that there is plenty of room for confusion and disagreement over the course of development and over what measures are desirable to control or encourage it.

Over time, rather than a regular or continuous advance, there are surges forward followed by standstills or periods of little change. This process is best observed in respect of the historical record, and was elaborated on in the case of the UK, where individual waves of development were associated with particular technologies being adopted, involving resources such as the iron ore and coalfields of Wales, northern England and Scotland. In countries that have undergone recent economic development, these waves are less observable in the historical record because the summed experience of the advanced countries is transferred to them wholesale, and irrelevant or outmoded technologies are largely bypassed. This is the case for the NICs, where the "success story" is partly the story of a sudden acquisition of accumulated skills and methods from other parts of the world.

In the spatial dimension there is also great irregularity, and this is observable both in terms of change from reliance on one spatially located resource to another, and in terms of concentrations at particular centres where the benefits of urban agglomeration are found. Taking the British example, the spatial pattern of development in the early Industrial Revolution focused on northern Britain, where the Lancashire and Yorkshire industrial towns, and those of the central Scotland belt between Glasgow and Edinburgh, were centres of innovation and the sites for some of the raw materials. In the twentieth century, the resources have mostly been human rather than material, and the new focus has been in London and the southeast of England. The movement from heavy engineering and textiles, to consumer durables such as cars and electrical goods, and from these on to services of all kinds, meant a massive move from northern regions to the south.

The other kind of spatial irregularity is the concentration of development in certain urban centres. Some advanced countries, it is true, do not depend heavily on urban centres and their industries, and have, for example, an advanced agriculture which provides them with wealth. Countries such as New Zealand come into this category, but there are few such countries, and the standard case is that of a country whose industries are heavily concentrated in urban centres, often in just one or two cities. This is true both for advanced countries and for the LDCs, where rapid migration out of the countryside and into the towns is in progress.

It is sometimes argued that the new information technologies allow society to avoid the old concentrations, and that part of counter-urbanization is to do with a new technology whereby very diffuse patterns are possible – teleworking, deconcentration of manufacturing processes to distant regions, all linked in through modern communications to a remote centre. Evidence from recent industrial history, however (Malecki 1991, Storper 1991), and from the data gathered in Chapter 3, is that concentrations are just as important as ever, although they may be new ones and in new locations.

Because of this continued tendency to concentrate, there are also tendencies to inequality, in wealth and income levels, because the workers in the new concentrated industries are commonly paid more for their special skills and for their scarce products than workers in other regions. What has been debated over the years is the value of intervention in order to overcome the inequalities, to make income levels approximately even between different industries and regions.

Intervention

A constant concern of students of development is whether state intervention to promote development, or intervention by any outside body, is justified or effective. The discussion in Chapter 2 suggested a major division of thought between a right- and left-wing view of this matter. In that chapter, the discussion was restricted to state interventions within the confines of a single nation state. The argument could have been enlarged to include international aid and interventions, and here there has been a vigorous debate, ending with the substantive defeat of the idea and the practice of development aid, despite the caricature of the opponents given by Toye (1987).

Most of the intra-nation debate was about the value of intervention in order to reduce the inequality thrown up by the development process, as innovations brought an advantage to specific social groups or geographical regions and countries.

In the last century, a number of writers like William Morris (MacCarthy 1994) or Robert Owen, followed into the early part of this century by Ebenezer Howard (Moss-Eccardt 1973), saw the need to correct a growing urban–rural imbalance, to which newly developing industries contributed. Through most of the twentieth century, a growing consensus of opinion was that intervention was indeed justified, and the central aim of regional development policy, in the UK and elsewhere, was constructed as the reduction of the inequalities through policies to attract industry, through projects to build new infrastructure such as roads or hydroelectric installations, or through national sectoral policies which provided public services at equal prices regardless of the cost of bringing them to the region. Part of the welfare state philosophy was the provision of services to all at an even level of supply, and regional development thus became part of the welfare idea. This was challenged in the 1980s by the neo-liberal reversion to markets and freer competition, which implied no special treatment for regions with poorer service provision.

Intervention was increasingly criticized in the advanced economies, because of the very telling evidence that it was unsuccessful in reducing inequalities, as in the UK over the past 50 years, and because its strategy was often wrong. In the British case, as in Spain and Europe generally, this intervention had been to maintain specific regions and industries despite global tendencies for these industries to be relocated in newly developing countries with major advantages such as lower material and labour costs. Intervention to support steelmaking in Scotland was misplaced philanthropy, because it consigned the central Scotland belt to a sector with poor growth and profitability in the whole of Europe, and reliant on resources that were becoming scarcer and more costly in the home region. Eventual conversion has cost the central government more in redirecting aid to new industries and locations, and has ensured high unemployment and inappropriate training for large numbers of workers in the declining industries of coal, steel and heavy engineering. The debate on intervention was thus apparently won by those denying its utility.

In the LDCs, too, there were plenty of examples of state intervention to help regions, which were misplaced from the point of view of promotion of development. In Latin America, for example (Morris 1987), intervention was sometimes fighting the results of natural disaster, sometimes defending political frontiers,

175

and sometimes building monuments to dictators. Rarely was it policy to help local people. From the command economies of China and Russia (Ch. 7), the role of regional development is seen to be part of national strategy, attempting to integrate distant regions into the centre. In India (Ch. 7) there were efforts to spread development initiatives democratically, but this, in the context of deep poverty and an expanding population, meant an ineffective weak gruel of aid to too many regions.

The long debate over intervention missed out, however, a vital role for state intervention to help regions or industries. As pointed out in Chapter 8, a major success of the NICS has been in intervening, not to bolster up declining regions, but to support new industries in the areas chosen by these industries, through giving them export credits and designating them as specially protected industries through the early stages of their growth. Intervention of this kind has positive results, and leads to global competitiveness for these industries, allowing them to kindle further growth in related industries within the country. There can be a regional aspect to such policy, although not strongly exemplified in the countries studied here, in moves to help regions that are responsible for the designated growth industries.

There is always a need for care in such policies of intervention, and in the case of the NICS, the protection and promotion was gradually reduced in the successful industries, so that they remained as strong exporters to the rest of the world based on their own efficiency. Early protection is followed by later opening to the winds of competition.

The state is also justified in intervention in another way, in preparing a level playing-field for a following development thrust. This could be through a sectoral policy, such as that for agrarian reform as in Taiwan, or through the take-over of domestic industries controlled by outside interests or by domestic monopolists who run these industries inefficiently, as in most South American countries, or in Russia at the time of the Communist Revolution of 1917. Temporary state interventions of this kind, if they are not followed by too extensive a period of state control as in the Russian case, may be beneficial in allowing a new start, with new social actors. Beyond this, the state may be able to help regions, not through regional or sectoral policy, but through macroeconomic policies, such as the maintenance of a stable currency, a measure that reduces the risks to entrepreneurs investing in new industries or new regions.

In summary, the case to be made is not totally for or against state intervention. Instead, intervention needs to be selective, in time and space, and to be able to retreat, either when statism threatens the strength of civil society so that there is no further innovation or enterprise (Russia), or when the competitiveness of industries is likely to be reduced through the creation of an artificially protected environment, as in the UK or in India since the Second World War.

Spatial development

For geographers, a question with regard to development must always be whether they hold some special key that could be used to plan development formally and which would help regions. Spatial planning reached its apogee in the 1960s, as an interpretation of Perroux's work, and this approach was found to be wanting both in theory and practice (Ch. 2). In effect, it was a misinterpretation which was made, and Perroux's central point was missed. This had been that polarization, concentration on single points, is a universally observable phenomenon, and a beneficial one, although not a planning doctrine. Rather than encase development in a spatial straitjacket, the need for geographers and others was, and remains, to understand the polarization process in space, and perhaps to be able to anticipate it.

In other words, there are spatial patterns, and they are worth observing and taking into account. But to set up spatial patterns as a separate exercise, and plant economic activities within them, is fraught with dangers because of the complexity of spatial phenomena, and the results show that such an approach is unworkable. Instead, there is value in observing, at various scales, the spatial patterns and their dynamic; at the international level, the spread of the development idea amongst the East Asian countries is one large-scale pattern. At a national level, there are the patterns observable over time and space in any country, made up of a multiple overlay of past geographies, to be set beside current economic activities. At the urban and regional level, there are the relations between urban industrial and service concentrations, and the activities in the surrounding countryside.

Key factors in development

Many writers have made their primary focus the identification of the key factor, the basic cause that has permitted some countries to progress more than others. This book has not attempted to join in the debate, but it is worth noting where we stand following the discussion in previous chapters. Classical economics considered only three factors in production: land, labour and capital. This was simple enough and made simpler by considering the factors in a more or less standardized fashion. In the eighteenth century Adam Smith, and following him, Ricardo in the nineteenth century, had considered agriculture, or land, to be of fundamental importance for the increase of wealth (Chisholm 1980, 1982). Later in the century, more attention was given to the factor of capital, seen by both Marx and by the writers who supported capitalism as the key; labour was also acknowledged to be important. However, this was usually undifferentiated labour, taking no cognizance of special skills or abilities (i.e. the quality of labour).

In the present century, writers such as Schumpeter (1939) have called attention to the role of technological innovation. Those regions and industries with command over new technologies, especially where it is possible to limit the spread of the special knowledge to other regions and industries, have been great sources of profit and wealth. This point was picked up in Chapter 3, which focused on manufacturing industry. Some of the innovations, such as harvesting and ploughing machinery, have been used in agriculture, but much of the actual innovation may be regarded as happening in manufacturing industry, and its benefits handed on to other sectors.

Certainly, technology is a major driving force, and responsible for much of the long waves that have been observed in Britain (Ch. 6) and elsewhere, the changes causing massive shifts in the style and level of regional development, as well as pushing whole nations to the forefront of economic development. Technology has meant that land (or resources, if we expand the term "land" to include all natural elements used by a population) has not had a single meaning or importance, but one that varies with the state and the type of technology in use. Resources can be exhausted, but they are often replaced by others before total exhaustion sets in.

However, technology does not provide a complete explanation, or even a dominant one in the way argued by Malecki (1991). Instead, we have argued here for the importance of a combination of factors unique to each country, and a unique historical sequence of discovery, use of different factors, and control systems within a nation. One of the overlooked elements in this complex, at least as far as economists are concerned, has been the social structure.

The social context

Acknowledgement given to the question of social structure (Ch. 4) as a factor in development is a recent matter from the point of view of development studies, and for many may seem an unsettling feature, taking development out of the limited field of economics. Consideration of social aims as part of the aims of development itself is also unsettling, as it broadens out the study considerably. Material wealth of a fairly simple kind may be the aim of people in the early stages of development, but this certainly changes over time, or at least the non-material aims are given public recognition as development goes on. Our most sophisticated societies would seem to have major aims connected with the position of the individual in society, for public acceptance or acceptance within a selected group. Perhaps these aims are always with us, as innate needs of the human being. That they come to light in advanced societies may merely represent the breakdown of primitive societies where a multiplicity of human aims were satisfied within the limits of a small community.

Another aspect of the social that emerges as important is its role in helping the productive economy. Once we move beyond the consideration of humanity as merely numbers, the presence of special skills and abilities can be recognized, so that education and training become not just a welfare measure, but an investment in development.

In addition, at various points the way in which societies are organized has been mentioned as a feature helping or hindering their development. A strong "civil society" is increasingly being brought into the equation as an explanatory factor, and this term is used to refer to all kinds of organization below the state level, including firms, clubs, the family, and other groupings of formal and informal nature. Where the main levels of organization are restricted to the state and the individual, there is a partial vacuum as far as the organization of economic activities is concerned, and this is a poor base for development (Fukuyama 1995). Part of the explanation of success in the USA, historically, and in Japan and the Far East today, is to be found in the strength of civil society, according to Fukuyama.

What the social concerns do present to the geographer is a role for a social scientist who is not confined by sectoral limitations, and who may be trained to take into account both society and economy. In other words, the very breadth of the development theme presents a special opportunity to geographers. Both in the area of human aims and priorities, and in the matter of identifying and examining civil society, there are areas of research available to this discipline.

And finally ...

In Chapter 9, another kind of case for the geographer was made out. Part of the endowment of the current vogue for post-modernist thinking is that people and places are different, and should be treated as such, rather than as examples of a type. With reference to regions, this means that there is a role to play in the administration of plans for regions in terms of their special characteristics and the linkages between these characteristics. This was the classical area of expertise of regional geographers up to the 1950s, except that they were often inclined to classify the regions into types. Now there is some recognition that each region does have a special mix of economy, society and environment, and in each region a separate cultural history is to be sought.

These differences can be aligned with differences in aims for development between regions, so that planning for regions in their own right can be undertaken, combining knowledge of special problems and special aims of the population, and linking all of these to possible plans or policies for the future. To consider the political conditions under which such local decision-making is made possible goes beyond the frame of reference of this book.

Bibliography

Abe, H. 1996. New directions for regional development planning in Japan. See Alden & Boland (1996), 273-95.

Aberbach, J.D., D. Dollar, K.L. Sokoloff 1994. *The role of the state in Taiwan's development*. New York: M.E. Sharpe.

Agarwala, A.N. & S.P. Singh 1958. *The economics of underdevelopment*. Oxford: Oxford University Press.

Alden, J. & P. Boland 1996. *Regional development strategies: a European perspective*. London: Jessica Kingsley.

Allen, J. 1995. Crossing borders: footloose multinationals. In *A shrinking world: global unevenness and inequality*, J. Allen & C. Hamnett (eds), 55–102. Oxford: Oxford University Press.

Alonso, W. 1980. Five bell shapes in development. *Papers, Regional Science Association* **45**, 5–16.

Amin, A. 1989. A model of the small firm in Italy. See Goodman & Bamford (1989), 111 20.

—— & N. Thrift (eds) 1994. *Globalization, institutions, and regional development in Europe*. Oxford: Oxford University Press.

Arnold, U. & K.N. Bernard 1989. Just in time: some marketing issues raised by a popular concept in production and distribution. *Technovation* **9**, 401–31.

Auty, R.M. 1990. The impact of heavy-industry growth poles on South Korean spatial structure. *Geoforum* **21**, 23–33.

—— 1994. Industrial policy reform in six large newly industrializing countries: the resource curse thesis. *World Development* **22**, 11–26.

Balchin, P.N. 1990. *Regional policy in Britain; the north–south divide*. London: Paul Chapman.

Banco Bilbao Vizcaya 1991. *Renta Nacional de España y su distribución provincial*. Bilbao: BBV.

Barbancho, A. 1979. *Disparidades regionales y ordenación del territorio*. Barcelona: Ariel.

Barke, M. 1989. Provincial disparities in levels of living: the case of Spain 1970–81. *Geojournal* **18**, 407–16.

—— & G. O'Hare 1991. *The Third World: diversity, change and interdependence*, 2nd edn. Harlow: Longman.

—— & G. Park 1994. *Spatial patterns of levels of living: Spain 1970–1989*. Occasional Papers, New Series No. 9. Division of Geography and Environmental Management, University of Northumbria.

Barquero, A.V. & M. Hebbert 1985. Spain: economy and state in transition. In *Uneven development in southern Europe*, R. Hudson and J. Lewis (eds), 284–308. London: Methuen.

181

Beenstock, M. 1983. *The world economy in transition*. London: Allen & Unwin.

Bhagwati, J. 1995. *India in transition; freeing the economy*. Oxford: Clarendon.

Birdsall, N., D. Ross, R. Sabot 1995. Inequality and growth reconsidered: lessons from East Asia. *World Bank Economic Review* **9**, 477–508.

Bleitrach, D. & A. Chenu 1982. Regional planning; regulation or deepening of social contradictions? The example of Fos-sur-Mer and the Marseilles Metropolitan Area. In *Regional Planning in Europe*, R. Hudson & J. Lewis (eds), 148–77. London: Pion.

Bobek, H. 1974. Zum Konzept des Rentenkapitalismus. *Tijdschrift voor Economische en Sociale Geographie* **65**, 73–8.

Boudeville, J.-R. 1966. *Problems of regional economic planning*. Edinburgh: Edinburgh University Press.

Bradshaw, M.J. (ed.) 1991. *The Soviet Union: a new regional geography?* London: Belhaven.

—— (ed.) 1997. *Geography and transition in the post-Soviet republics*. Chichester: John Wiley.

Brass, P.R. 1991. *The politics of India since independence*. Cambridge: Cambridge University Press.

Browder, J.O. 1989. *Fragile lands of Latin America: strategies for sustainable development*. Boulder: Westview.

Campbell, A., C. Keen, G. Norman, R. Oakeshott 1977. *Worker-owners: the Mondragon achievement*. London: Anglo-German Foundation for the Study of Industrial Society.

Carreras, A. 1990. Cataluña, primera región industrial de España. See Nadal & Carreras (1990), 259–95.

Castells, M., A. Portes, L. Benton 1989. *The informal economy: studies in advanced and less developed countries*. Baltimore: Johns Hopkins University Press.

Chisholm, M. 1980. The wealth of nations. *Transactions of the Institute of British Geographers* **5**, 255–76.

—— 1982. *Modern world development: a geographical perspective*. London: Hutchinson.

—— 1990. *Regions in recession and resurgence*. London: Unwin Hyman.

Chu, Y. 1994. The state and the development of the automobile industry in South Korea and Taiwan. See Aberbach et al. (1994), 125–69.

Clem, R.S. 1997. The new Central Asia: prospects for development. See Bradshaw (1997), 165–85.

Clout, H.D. 1982. *The geography of post-war France*. Oxford: Pergamon.

Cole, J.P. 1987. Regional inequalities in the Peoples' Republic of China. *Tijdschrift voor Economische en Sociale Geografie* **78**, 201–13.

Conroy, M. 1973. Rejection of growth center strategy in Latin American regional development planning. *Land Economics* **49**, 371–80.

Cooke, P. & K. Morgan 1994. Growth regions under duress: renewal strategies in Baden Wurttemberg and Emilia-Romagna. In *Globalization, institutions and regional development in Europe*, A. Ash & N. Thrift (eds), 91–117. Oxford: Oxford University Press.

Corbridge, S. 1986. *Capitalist world development: a critique of radical development geography*. London: Macmillan.

Cowen, M.P. & R.W. Shenton 1996. *Doctrines of development*. London: Routledge.

Cutter, S.L. 1986. Changes in interstate rankings 1931–1980. *Geographical Review* **76**, 276–87.

Daewood, D. & Sjafrizal 1991. Aceh: the LNG boom and enclave development. See Hill (1991), 107–24.

Damesick, P. 1987. The changing economic context for regional development in the UK. In

Regional problems, problem regions and public policy in the UK, P. Damesick & P. Wood (eds), 1–17. Oxford: Clarendon.

Dietz, J.L. (ed.) 1995. *Latin America's economic development: confronting crisis.* Boulder: Lynne Riener.

Dmitrieva, O. 1996. *Regional development: the USSR and after.* London: UCL Press.

Drucker, P. 1985. *Innovation and entrepreneurship: practice and principles.* London: Heinemann.

—— 1989. *The new realities.* Oxford: Heinemann.

Dutt, A.K. & M.M. Geib 1987. *Atlas of South Asia.* Boulder: Westview.

Edwards, C. 1992. Industrialization in South Korea. See T. Hewitt et al. (1992), 97–127.

Escobar, A. 1995. *Encountering development: the making and unmaking of the Third World.* Princeton, NJ: Princeton University Press.

Escude, C. 1987. *Patología del nacionalismo: el caso argentino.* Buenos Aires: Editorial Thesis.

Farmer, B.H. 1993. *An introduction to South Asia,* 2nd edn. London: Routledge.

Findlay, A.M., F.L.N. Li, A.J. Jowett & R. Skeldon 1996. Skilled international migration and the global city: a study of expatriates in Hong Kong. *Transactions of the Institute of British Geographers* 21, 49–61.

Frank, A.G. 1967. *Capitalism and underdevelopment in Latin America.* London: Monthly Review Press.

Friedmann, J. 1966. *Regional development policy: a case study of Venezuela.* Cambridge, MA: MIT Press.

—— 1986. The world city hypothesis. *Development and Change* 17, 69–83.

—— 1988. *Life space and economic space: essays in Third World planning.* New Brunswick: Transaction Books.

—— & C. Weaver 1979. *Territory and function: the evolution of regional planning.* London: Edward Arnold.

Fröbel, F., J. Heinrichs, O. Kreye 1980. *The new international division of labour.* Cambridge: Cambridge University Press.

Fukuyama, F. 1995. *Trust: the social virtues and the creation of prosperity.* London: Hamish Hamilton.

Garcia Delgado, J.L. 1991. *España, economía.* Madrid: Espasa Calpe.

Geertz, C. 1963. *Agricultural involution: the process of ecological change in Indonesia.* Berkeley: University of California.

Glickman, N. & A. Glasmeier 1989. The international economy and the American South. See Rodwin & Sazanami (1991), 60–80.

Gligo, N. 1993. Environment and natural resources in Latin American development. See Sunkel (1993), 185–222.

Gonzalez Casanova P. 1964/65. Internal colonialism and national development. *Studies in International Comparative Development* 1, 27–37.

—— 1969. International colonialism and national development. In *Latin American radicalism,* I.L. Horowitz et al. (eds). New York: Vintage.

Goodman, E. & J. Bamford (eds) 1989. *Small firms and industrial districts in Italy.* London: Routledge.

Gore, C. 1984. *Regions in question: space, development theory, and regional policy.* London: Methuen.

Government Information Office of Taiwan 1993. *The Republic of China.* Taipeh, Taiwan.

Grabher, G. (ed.) 1993. *The embedded firm: on the socio-economics of industrial networks.* London: Routledge.

Granovetter, M. 1985. Economic action and social structure; the problem of embeddedness. *American Journal of Sociology* **91**, 481–510.

Gravier, J.F. 1947. *Paris et le désert français*. Paris: Flammarion.

—— 1970. *La question régionale*. Paris: Flammarion.

Gaudemar, J-P. & R. Prud'homme 1991. Spatial impacts of deindustrialization in France. See Rodwin & Sazanami (1991), 105–36.

Gwynne, R. 1992. *New horizons? Third World industrialization in an international framework*. London: Longman.

Hall, P. 1991. Structural transformation in the regions of the UK. See Rodwin & Sazanami (1991), 39–69.

—— M. Breheny, R. McQuaid, D. Hart 1987. *Western sunrise: the genesis and growth of Britain's major high tech corridor*. London: Allen & Unwin.

—— & P. Preston 1988. *The carrier wave: new information technology and the geography of innovation 1846–2003*. London: Unwin Hyman.

Hansen, N.M. 1968. *French regional planning*. Edinburgh: Edinburgh University Press.

Hardjono, J. 1994. Resource utilization and the environment. See Hill (1994), 179–215.

Harrison, B. 1992. Industrial districts: old wine in new bottles? *Regional Studies* **26**, 469–83.

Hartshorne, R. 1939. *The nature of geography: a critical survey of current thought in the light of the past*. Lancaster, PA: Association of American Geographers.

Hayek, F. 1986. *The road to serfdom*. London: Ark.

Hennessy, A. 1978. *The frontier in Latin American history*. London: Edward Arnold.

Herd, R. & R. Jones 1994. Spotlight on Korea. *OECD Observer* **188**, 1–6.

Hettne, B. 1990. *Development theory and the three worlds*. London: Longman.

Hewitt, T., H. Johnson, D. Wield (eds) 1992. *Industrialization and development*. Oxford: Oxford University Press.

Higgins, B. & D.J. Savoie (eds) 1988. *Regional economic development: essays in honour of Francois Perroux*. Boston: Unwin Hyman.

Hill, H. (ed.) 1991. *Unity and diversity: regional economic development in Indonesia since 1970*. Oxford: Oxford University Press.

—— (ed.) 1994. *Indonesia's new order: the dynamics of socio-economic transformation*. Sydney: Allen & Unwin.

Hirschman, A.O. 1958. *The strategy of economic growth*. New Haven, CT: Yale University Press.

Holland, S. 1976. *Capital versus the regions*. London: Macmillan.

Hsieh, C. 1964. *Taiwan – Ilha Formosa*. London: Butterworths.

Huang, S.W. 1993. Structural change in Taiwan's agricultural economy. *Economic development and cultural change* **42**, 43–65.

Hudson, R. 1986. Creating an industrial wasteland: capital, labour and the state in Northeast England. See Martin & Rowthorn (1986), 169–213.

Innis, H. 1930. *The fur trade in Canada*. New Haven, CT: Yale University Press.

Johnston, RJ., P. Taylor & M.J. Watts 1995. *Geographies of global change: remapping the world in the late twentieth century*. Oxford: Blackwell.

Kaiser, R.J. 1991. Nationalism: the challenge to Soviet federalism. See Bradshaw (1991), 39–66.

—— 1997. Nationalism and identity. See Bradshaw (1997), 9–30.

King, A.D. 1990. *Global cities: post-imperialism and the internationalization of London*. London: Routledge.

Knox, P.L. & J. Agnew 1989. *The geography of the world economy*. London: Edward Arnold.

Lacoste, Y. 1975. *Géographie du sous-développement.* Paris: Presses Universitaires de France.

Lau, L.J. (ed) 1990. *Models of development: a comparative study of growth in South Korea and Taiwan.* San Francisco, CA: ICS Press.

Le Grand, J. 1982. *The strategy of equality: redistribution and the social services.* London: Allen & Unwin.

Lewis, R.A. 1992. *Geographic perspectives on Soviet Central Asia.* London: Routledge.

Liebowitz, R.D. 1991. Spatial inequality under Gorbachev. See Bradshaw (1991), 15–38.

—— 1992. Soviet geographical imbalances and Soviet Central Asia. See Lewis (1992), 101–31.

Lipton, M. & R. Longhurst 1989. *New Seeds and poor people.* London: Unwin Hyman.

Lo, F. & K. Salih 1981. Growth poles, agropolitan development, and polarization reversal: the debate and search for alternatives. See Stöhr & Taylor (1981), 123–52.

Lopes, J.B. 1977. Développement capitaliste et structure agraire au Brésil. *Sociologie du Travail* **19**, 59–71.

Lyons, T.P. 1990. Interprovincial disparities in China: output and consumption, 1952–1987. *Economic Development and Cultural Change* **39**, 471–505.

MacCarthy, F. 1994. *William Morris: a life for our time.* London: Faber & Faber.

Mair, A. 1995. *Honda's global local corporation.* New York: St Martin's Press.

Malecki, E.J. 1991. *Technology and economic development: the dynamics of local, regional, and national change.* London: Longman Scientific.

Markusen, A. 1985. *Profit cycles, oligopoly and regional development.* Cambridge, MA: MIT Press.

Marshall, J.N., et al. 1988. *Services and uneven development.* Oxford: Oxford University Press.

Marsland, D. 1996. *Welfare or welfare state? Contradictions and dilemmas in social policy.* London: Macmillan.

Martin, R. 1986. Thatcherism and Britain's industrial landscape. See Martin & Rowthorn (1986), 238–90.

—— & B. Rowthorn (eds) 1986. *The geography of de-industrialization.* London: Macmillan.

Martinez, J. & A. Diaz 1996. *Chile: the great transformation.* Washington: Brookings Institute and UNRISD.

Maslow, A.H. 1954. *Motivation and personality.* New York: Harper.

Massey, D. 1984. *Spatial divisions of labour: social structures and the geography of production.* London: Macmillan.

Maxcy, G. 1981. *The multinational motor industry.* London: Croom Helm.

McLuhan, M. 1964. *Understanding media: extensions of man.* London: Routledge & Kegan Paul.

Mendez Gutierrez del Valle 1990. *Geografía de España: las actividades industriales.* Madrid: Sintesis.

Morris, A.S. 1972. The regional problem in Argentine development. *Geography* **57**, 289–306.

—— 1981. *Latin America: economic development and regional differentiation.* London: Hutchinsons.

—— 1987. Regional development: trends and policies. In *Latin American development: geographical perspectives,* D. Preston (ed.), 141–68. London: Longman.

—— 1992a. Neoconservative policies in Argentina and the decentralization of industry. In *Decentralization in Latin America,* A. S. Morrris & S. Lowder (eds), 77–92. New York: Praeger.

—— 1992b. Spain's new economic geography: the Mediterranean axis. *Scottish Geographical Magazine* **108**, 92–8.

—— 1996a. Geopolitics in South America. In *Latin American development: geographical perspectives*, 2nd edn, D. Preston (ed.), 272–94. London: Longman.

—— 1996b. Regional industrial promotion in Argentine Patagonia: end of an era. *Tijdschrift voor economische en sociale geografie* **87**, 399–406.

—— & S. Lowder 1992. Flexible specialization: the application of theory in a poor-country context: Leon, Mexico. *International Journal of Urban and Regional Research* **16**, 190–201.

Moss-Eccardt, J. 1973. *Ebenezer Howard: an illustrated life of Sir Ebenezer Howard, 1850–1928*. Aylesbury: Shire.

Myers, R.H. 1990. The economic development of the Republic of China on Taiwan, 1965–81. See Lau (1990), 17–63.

Myrdal, G. 1957. *Economic theory and underdeveloped regions*. London: Duckworth.

—— 1977. *Asian drama: an enquiry in the poverty of nations*. London: Penguin.

Nadal, J. & A. Carreras 1990. *Pautas regionales de la industrializacion española*. Barcelona: Ariel.

Naylon, J. 1992. Ascent and decline in the Spanish regional system. *Geography* **77**, 46–62.

North, D.C. 1961. *The economic growth of the United States*. Englewood Cliffs, NJ: Prentice Hall.

Ohmae, K. 1995. *The end of the nation-state: the rise of regional economies*. New York: Free Press.

Ozornoy, G.I. 1990. Some issues of regional inequality in the USSR under Gorbachev. *Regional Studies* **25**, 381–391.

Paauw, D. & J. Fei 1975. *The transition in open dualistic economies: theory and Southeast Asian experience*. New Haven, CT: Yale University Press.

Paddison, R. 1993. City marketing, image reconstruction and urban regeneration. *Urban Studies* **30**, 339–350.

Park, E.Y. 1992. Economic transformation of Asia-Pacific region and Korea. In *Korea in a turbulent world*, I.Y. Chung (ed.), 355–95. Seoul: Sejong Institute.

Peck, J. 1994. Regulating labour: the social regulation and reproduction of local labour-markets. See Amin & Thrift (1994), 147–76.

Peet, R. & M. Watts 1993. Introduction: development theory and environment in an age of market triumphalism. *Geography* **69**, 227–53.

Perroux, F. 1955. Note sur la notion de pole de croissance. *Economie Appliqué* **8**, 307–20; see also I. Livingstone (ed.) 1971. *Economic policy for development* (London: Penguin) for a translation.

—— 1988. The pole of development's new place in a general theory of economic activity. See Higgins & Savoie (1988), 48–76.

Pinch, S. 1985. *Cities and services: the geography of collective consumption*. London: Routledge & Kegan Paul.

Piore, M. & C. Sabel 1984. *The second industrial divide: possibilities for prosperity*. New York: Basic Books.

Polanyi, K. 1957a. *The great transformation*. Boston: Beacon.

—— 1957b. The economy as instituted process. In *Trade and markets in the early empires*, C.W. Arensberg & H.W. Pearson (eds). New York: Free Press.

—— 1977. *The livelihood of man*. New York: Academic Press.

Powell, W.W. 1990. Neither market nor hierarchy: network forms of organization. *Research in Organizational Behavior* **12**, 295–336.

Prestwich, R. & P. Taylor 1990. *Introduction to regional and urban policy in the United Kingdom*. London: Longman.

Pryde, P.R. 1997. The post-Soviet environment. See Bradshaw (1997), 131–44.
Ranis, G. 1995. Another look at the East Asian miracle. *World Bank Economic Review* **9**, 509–34.
Redclift, M. 1987. *Sustainable development: red or green alternatives*. London: Methuen.
Reilly, C.A. 1995. *New paths to democratic development in Latin America: the rise of NGO-municipal collaboration*. Boulder: Lynne Riener.
Reitsma, H.J.A. 1982. Development geography, dependency relations and the capitalist scapegoat. *Professional Geographer* **34**, 125–30.
—— & J.M.G. Kleinpenning 1985. *The third world in perspective*. Assen: Van Gorcum.
Richardson, H. 1973. *Regional growth theory*. London: Macmillan.
—— 1975. *Regional development policy and planning in Spain*. Farnborough: Saxon House.
Riddell, R.B. 1981. *Ecodevelopment: economics, ecology and development*. London: Gower.
Rodwin, L. 1989. Deindustrialization and regional economic transformation. See Rodwin & Sazanami (1989), 3–25.
—— 1991. European industrial change and regional economic transformation: an overview of recent experience. See Rodwin & Sazanami (1991), 3–36.
—— & H. Sazanami (eds) 1989. *Deindustrialization and regional economic transformation: the experience of the United States*. Boston: Unwin Hyman.
—— & —— (eds) 1991. *Industrial change and regional economic transformation*. London: HarperCollins Academic.
Rogerson, R., A. Findlay, M. Coombes, A. Morris 1989. Measuring quality of life: some methodological issues. *Environment and Planning A* **21**, 1655–1666.
Rohwer, J. 1996. *Asia rising*. London: Nicholas Brealey.
Rondinelli, D.A. 1985. *Applied methods of regional analysis: the spatial dimensions of development policy*. Boulder: Westview.
—— & K. Ruddle 1978. *Urban and rural development: a spatial policy for equitable growth*. New York: Praeger.
Rostow, W.W. 1960. *The stages of economic growth: a non-Communist Manifesto*. Cambridge: Cambridge University Press.
Routledge, P. 1993. *Terrains of resistance: nonviolent social movements and the contestation of place in India*. Westport, CT: Praeger.
—— 1995. Resisting and reshaping the modern: social movements and the development process. See Johnston et al. (1995), 263–79.
Rowthorn, B. 1986. Deindustrialisation in Britain. See Martin & Rowthorn (1986), 1–30.
Rubenstein, J.M. 1986. Changing distribution of the American automobile industry. *Geographical Review* **76**, 288–300.

Sabel, C. 1996. *Local partnerships and social innovation: Ireland*. Paris: OECD.
Sassen, S. 1991. *The global city: New York, London, Tokyo*. Princeton, NJ: Princeton University Press.
Saxenian, A. 1994. *Regional advantage: culture and competition in Silicon Valley and Route 128*. Cambridge, MA: Harvard.
Schumpeter, J.A. 1939. *Business cycles: a theoretical and historical analysis of the capitalist process*. New York: McGraw-Hill.
Schvartzer, J. 1987. *Promoción industrial en Argentina: características, evolución y resultados*. Buenos Aires, Centro de Investigaciones sociales sobre el estado y la administración, Documentos del CISEA 90.
Scitovsky, T. 1990. Economic development in Taiwan and South Korea. See Lau (1990), 127–81.
Scottish Office 1992. *Scottish rural life: a socio-economic profile of rural Scotland*. Edinburgh: Scottish Office.

Selya, R.M. 1993. Economic restructuring and spatial changes in manufacturing in Taiwan, 1971–1986. *Geoforum* **24**, 115–26.

Sforzi, F. 1989. The geography of industrial districts in Italy. See Goodman & Bamford (1989), 153–73.

Shuurman, F.J. (ed.) 1993. *Beyond the impasse: new directions in development theory*. London: Zed Books.

Sinnott, P. 1992. The physical geography of Soviet Central Asia and the Aral Sea problem. See Lewis (1992), 74–97.

Slater, D. 1976. Underdevelopment and spatial inequality: approaches to the problems of regional planning in the Third World. *Progress in Planning* **4**, 97–167.

—— 1995. Trajectories of development theory: capitalism, socialism and beyond. See Johnston et al. (1995), 63–76.

Smith, D. 1989. *North and south: Britain's economic, social and political divide*. Harmondsworth: Penguin.

Smith, N. 1982. Theories of underdevelopment: a response to Reitsma. *Professional Geographer* **34**, 332–37.

Smith, P. 1992. Industrialization and environment. See Hewitt et al. (1992), 277–302.

Smout, T.C. 1986. *A century of the Scottish people*. Glasgow: Collins.

Stamp, L.D. & S.H. Beaver 1954. *The British Isles: a geographic and economic survey*. London: Longmans Green.

Stern, R.W. 1993. *Changing India*. Cambridge: Cambridge University Press.

Sternlieb, G. & J. Hughes 1978. *Revitalizing the Northeast*. New Brunswick: New Jersey State University.

Stöhr, W.B. 1981. Development from Below. See Stöhr & Taylor (1981), 39–72.

—— (ed.) 1990. *Global challenge and local response: initiatives for economic regeneration in contemporary Europe*. London: Mansell.

—— & D.R.F. Taylor (eds) 1981. *Development from above or below? The dialectics of regional planning in developing countries*. New York: John Wiley.

—— & F. Tödtling 1978. Spatial equity; some antitheses to current regional development strategy. *Papers, Regional Science Association* **38**, 33–53.

Storper, M. 1990. Industrialization and the regional question in the third world: lessons of postimperialism; prospects of post-Fordism. *International Journal of Urban and Regional Research* **14**, 423–44.

—— 1991. *Industrialization, economic development and the regional question in the Third World: from import substitution to flexible production*. London: Pion.

Street, J.H. 1995. The reality of power and the poverty of economic doctrine. See Dietz (1995), 27–44.

Sunkel, O. 1993. *Development from within: toward a neostructuralist approach for Latin America*. Boulder: Lynne Riener.

Taylor, D.R.F. & F. Mackenzie 1992. *Development from within*. London: Routledge.

Toffler, A. 1970. *Future shock*. London: Bodley Head.

Town and Country Planning Association 1987. *North–South divided: a new deal for Britain's regions*. London: TCPA.

Townsend, A. 1993. *Uneven regional change in Britain*. Cambridge: Cambridge University Press.

Toye, J. 1987. *Dilemmas of development*. Oxford: Basil Blackwell.

Turok, I. 1993. Contrasts in ownership and development; local versus global in "Silicon Glen". *Urban Studies* **30**, 365–86.

van Geenhuizen, M. & B. van der Knaap 1994. Dutch textile industry in a global economy. *Regional Studies* **28**, 695–711.

van Dijk, M.P. 1993. Industrial districts and urban economic development. *Third World Planning Review* **15**, 175–86.

Vasquez Barquero, A. 1986. El cambio del model de desarrollo regional y los nuevos procesos de difusion en España. *Estudios Territoriales* **20**, 87–110.

—— 1990. Local development initiatives under incipient regional autonomy: the Spanish experience in the 1980s. See Stöhr (1990), 354–74.

—— 1992. Local development and the regional state in Spain. In *Endogenous development and southern Europe*, G. Garofoli (ed.), 103–16. Aldershot: Avebury.

—— & M. Hebbert 1985. Spain: economy and state in transition. In *Uneven development in Southern Europe*, R. Hudson & J. Lewis (eds), 284–308. London: Methuen.

Vernon, R. 1966. International investment and international trade in the product cycle. *Quarterly Journal of Economics* **80**, 190–207.

—— 1979. The product cycle in a new international environment. *Oxford Bulletin of Economics and Statistics* **41**, 255–67.

Walton, J. 1975. Internal colonialism: problems of definition and measurement. In *Urbanisation and inequality*, W.A. Cornelius & F.M. Trueblood (eds), 29–50. London: Sage.

Weaver, C. 1984. *Regional development and the local community: planning, politics and social context*. Chichester: John Wiley.

Wilkinson, R.G. 1973. *Poverty and progress: an ecological model of economic development*. London: Methuen.

Williamson, J.G. 1965. Regional inequality and the process of national development: a description of the patterns. *Economic Development and Cultural Change* **13**, 3–45.

Williamson, O.E. 1975. *Markets and hierarchies: analysis and anti-trust implications*. New York: Free Press.

—— 1985. *The economic institutions of capitalism*. New York: Free Press.

World Bank 1975. *Rural development*. Washington, DC: World Bank.

World Commission on Environment and Development 1987. *Our common future* (Brundtland Report). Oxford: Oxford University Press.

Wyn Williams, S. 1977. Internal colonialism, core–periphery contrasts and devolution: an integrative comment. *Area* **9**, 272–8.

Index

Printed and bound by CPI Group (UK) Ltd, Croydon, CR0 4YY

23/10/2024

01777665-0010